Human Body

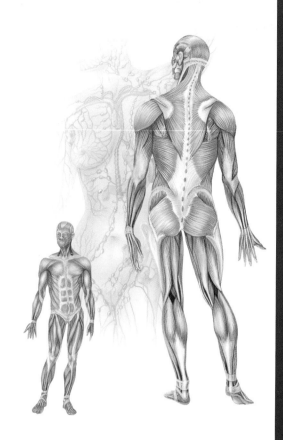

AUTHOR
Arco Editorial Team

EDITOR
Francisco Asesnsio Cerver

EDITORIAL MANAGEMENT
Nacho Asesnsio

MEDICAL CONSULTANTS
Dr. Antonio Paez and
Dr. Nuria Martinez
with the collaboration of
Dr. Evaristo Tardio Torio
for the chapter on Blood,
and Dr. Emilio Martin Orte
in relation to the illustrations

DESIGN AND LAYOUT
David Maynar

ILLUSTRATORS
Miquel Ferron
Miriam Ferron
Lidia di Blasi
Santiago Arcas
Francisco José Arcas
Francesc Florensa
Alex Vincon

©2000 for the English edition
Könemann Verlagsgesellschaft mbH
Bonnerstr. 126, D-50968 Cologne
Project coordination: Kristin Zeier
Production: Ursula Schümer
Printing and binding: EuroGrafica, Marano Vicenza
Printed in: Italy
ISBN 3-8290-2113-5

Translation from Spanish: Alayne Pullen and Simon Wiles in association with First Edition Translations Ltd, Cambridge, UK
Editing: Jean Macqueen in association with First Edition Translations Ltd, Cambridge, UK
Typesetting: The Write Idea in association with First Edition Translations Ltd, Cambridge, UK
Project Management: Andrew R. Davidson for First Edition Translations Ltd, Cambridge, UK

Table of Contents

Introduction 5

General structure of the human body 6
    Embryological origin 8
    Areas and regions 10

The Cell 13
The Skin 22
The Blood 28

The Systems 41
    Digestive system 42
    Nutrition 66
    Cardio-vascular system 76
    Respiratory system 90
    Muscular and skeletal system 108
    Reproductive system 116
    Nervous system 124
    Endocrine system 136
    Immune system 144
    Urinary system 152

Organs of Sense 162
    Sight 162
    Hearing 164
    Smell 166
    Taste 168
    Touch 170

Bibliography 172

Introduction

The general impression held by the general public whenever books about the human body or medicine are mentioned is of thick tomes and thousand of pages. This book is intended to break away from this image, while still aiming to provide a clear representation of the most important characteristics of the human body. The main intention, in fact, is to fulfil a primarily explanatory function.

The text covers the aspects related to the normal structure and functions of the body, while avoiding pathological or abnormal states. Accordingly, the areas on which most light is cast in this book are those of anatomy and physiology; in other words, bodily form and function respectively.

The layout of the book is intended to lead the reader from a description of the basic components of the organism, through to the way in which they are structured, and on to the complex associations of organs which go to make up the body's systems.

The opening chapters provide a general view of the structure of the human body, including aspects of embryo development. Chapters Three to Five deal with the basic elements of human architecture, the cells; the organ which gives the body its external appearance, the skin; and the vital fluid which occurs throughout the organism, the blood.

The last part of the book, and the most extensive, is the part which relates to the different systems, including those of which the systematization is considered very difficult, such as the immune system or the endocrine system. In this part, an overall anatomical description is provided for each system, as well as for the most important organs, and an explanation provided of the system's normal function.

The book combines a lucid text, with carefully researched explanations, with a large number of illustrations, and providing a good balance between them. It is not a book which will give the reader the feeling of being weighed down by a great mass of text, and to achieve this particular care has been given to the preparation of the illustrations.

We hope that this book will meet the needs of those who are looking for the basic facts relating to our bodies, by providing clear explanations and doing away with unnecessary complications.

Dr. Antonio Páez
Dr. Núria Martínez

General structure of the human body

Organization

1. Molecules
Combinations of atoms bonded together to form the basic building blocks of the human body. Examples include water, fats, carbohydrates, and proteins.

2. DNA
A combination of molecules called nucleotides (adenine, thymine, cytosine, guanine) which together make up a large macromolecule in the form of a double helix; this, with certain proteins called histones, forms the chromosomes, and contains all the genetic information required for the human body to be formed.

3. Mitochondria
These are where the process of cellular respiration takes place, including the metabolism of fats and sugars to obtain energy. They are the central energy source of the cell, essential for the cell's functioning.

4. Cell
The cell is the basic element of the various tissues. There are many types of cell: muscle cells, nerve cells, blood cells, and so on. Their shapes and sizes are very varied, reflecting the diversity of their functions within the organism. Certain cells have specialized functions, depending on the tissue of which they are a part.

5. Tissue
An assembly of cells. There are five basic types of tissue: epithelial tissue provides cover; connective tissue, which supports and gives form to the organs; muscular tissue, which produces movement; nervous tissue, which identifies sensations and is the seat of the higher functions; and blood and lymph tissue, which move substances around the body and defend against invasion.

6. Organ
An assembly of different types of tissue, which has a specific anatomical structure and a unique and irreplaceable function within the body.

7. System
An assembly of organs that function in a co-ordinated manner to fulfill a specific task. Examples include nutrition (the digestive system), support (the skeletal system), oxygenation (the respiratory system), the elimination of toxic substances (the excretory system), and so on.

8. Organism
The combination of all the systems makes up the organism. If each is functioning correctly, and if total coordination is maintained between them, then the complex mechanism that is the human body can be expected to function perfectly.

Organization

Structural organization of the human body

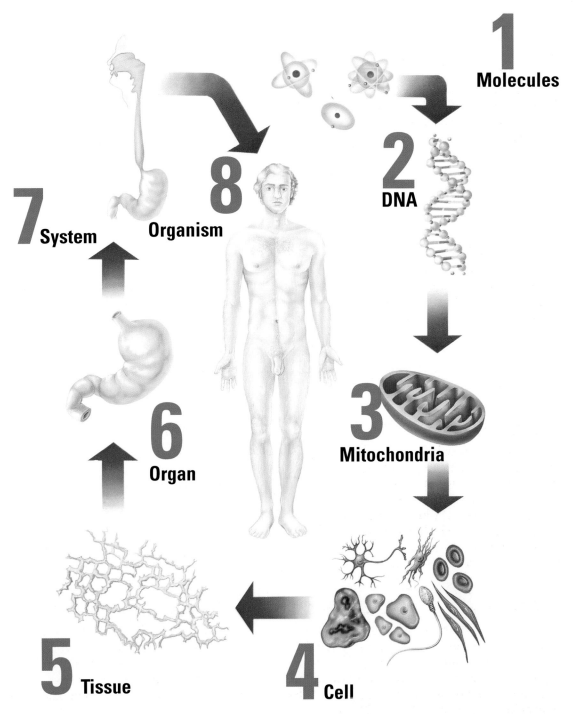

General structure of the human body
Organization

Embryological origin

The process of becoming a human being starts with fertilization, which consists of the union of two highly specialized cells, the male spermatozoon and the female ovum. This gives rise to the first cell of the new organism, the zygote. In the course of this process the sets of genetic material (the DNA helices) in the chromosomes of the maternal and paternal cells combine to form the DNA that is specific to the new cell.

This initial cell then divides repeatedly, resulting in a large number of daughter cells. These will eventually form the different tissues, organs, and systems that make up the new organism.

A few weeks after fertilization has occurred, the three basic layers of the future embryo have already formed: the ectoderm, mesoderm, and endoderm.

The outermost layer (the ectoderm) gives rise to the external layers of the skin and its appendages (such as hairs and sweat glands), the nervous system, the sensory epithelium of the sense organs, the pituitary gland, and the buccal cavity (the mouth) and its associated structures (teeth, salivary glands, and so on).

The intermediate layer (the mesoderm) develops to form the heart and the cardiovascular system, as well as muscles, bones, cartilages and joints, the kidneys and the rest of the urinary system, and the reproductive system.

The innermost layer (the endoderm) forms the digestive system and its associated organs (the liver and pancreas), the respiratory system, and certain internal glands such as the thymus, the thyroid, the parathyroid, and the tonsils.

Within the fully formed organism, all the elements and systems work in a coordinated and interlinked manner, with each one carrying out its own specific function.

A few weeks after fertilization, the three basic layers of the future embryo have already formed: the ectoderm, mesoderm, and endoderm.

Embryology

Cross-section through an embryo three weeks after fertilization, showing the ectoderm, mesoderm, and endoderm

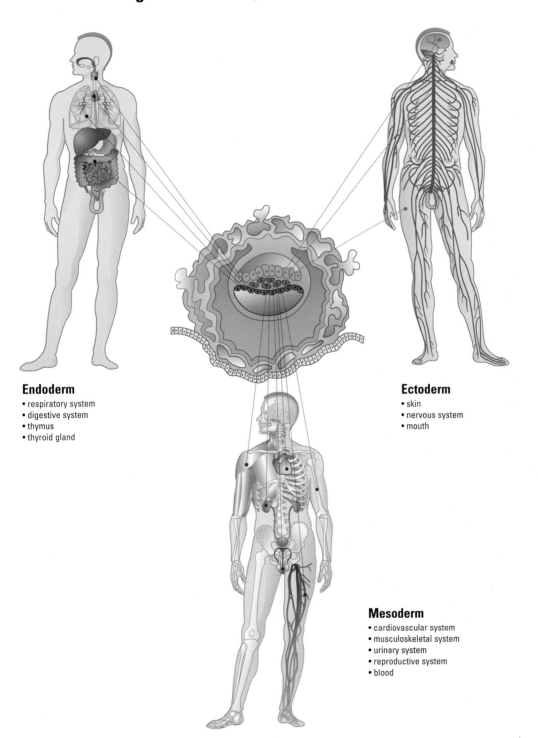

Endoderm
- respiratory system
- digestive system
- thymus
- thyroid gland

Ectoderm
- skin
- nervous system
- mouth

Mesoderm
- cardiovascular system
- musculoskeletal system
- urinary system
- reproductive system
- blood

General structure of the human body
Organization

Areas and regions of the human body

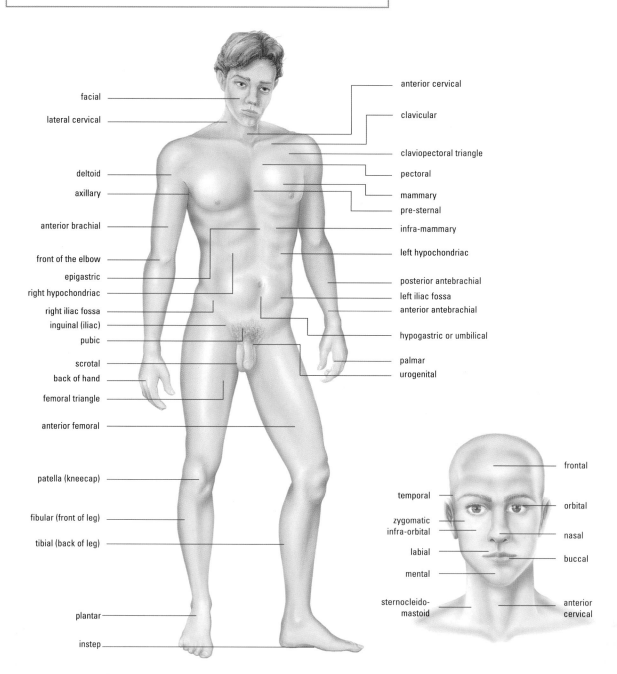

Areas/Regions

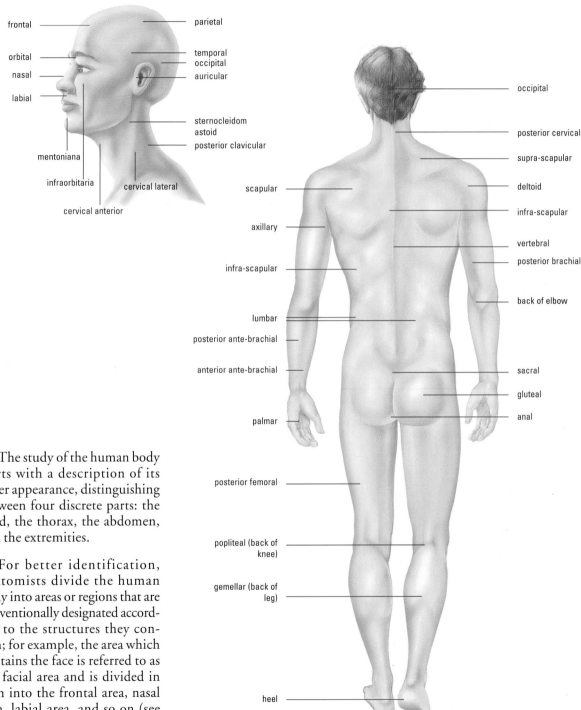

The study of the human body starts with a description of its outer appearance, distinguishing between four discrete parts: the head, the thorax, the abdomen, and the extremities.

For better identification, anatomists divide the human body into areas or regions that are conventionally designated according to the structures they contain; for example, the area which contains the face is referred to as the facial area and is divided in turn into the frontal area, nasal area, labial area, and so on (see illustrations).

Cells

The Cell

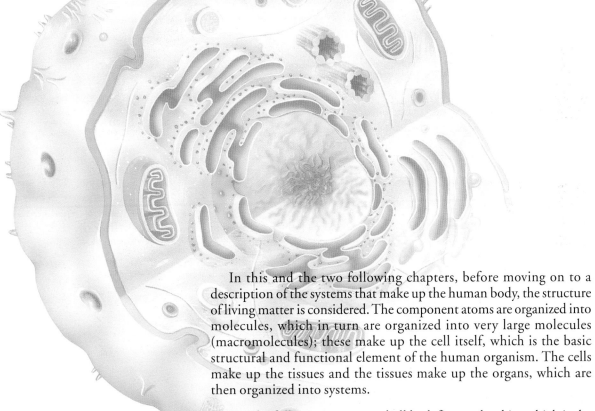

In this and the two following chapters, before moving on to a description of the systems that make up the human body, the structure of living matter is considered. The component atoms are organized into molecules, which in turn are organized into very large molecules (macromolecules); these make up the cell itself, which is the basic structural and functional element of the human organism. The cells make up the tissues and the tissues make up the organs, which are then organized into systems.

On the following pages we shall look first at the skin, which is the organ that is in contact with the outside world and forms our first line of defense to external attack. We shall be considering the external anatomical characteristics of the skin, as well as its microscopic internal structure.

At the end of this section, we shall describe the organ that extends all over the human body: the blood. We shall consider not only those aspects of the blood that are associated with circulation and the blood vessels (the peripheral blood), but also the processes by which it is formed.

Cells

The cell

Structure of living matter

The basic component of matter is the atom. The different kinds of atom make up the different elements (carbon, oxygen, hydrogen, etc.). Atoms join up to form molecules; one of the most important of these is water, which makes up 75% of the human body. The rest consists of organic compounds: the main ones are carbohydrates (the principal source of energy), the proteins (the structural elements of cells, which include the important compounds known as enzymes), and the lipids or fats (the structures and sources of energy reserves). The nucleic acids, deoxyribonucleic acid (DNA) and ribonucleic acid (RNA), play key parts in reproduction and inheritance.

Adenine Cytosine
Thymine Guanine

Helical structure of DNA with sequence of nucleotides.

From atoms to ions

Atoms (sodium and chlorine atoms are shown here as examples) maintain a balance between their electrons outside the nucleus and the protons within it. If an atom either loses or gains an electron the balance breaks down and the atom is converted into an ion, which has either a positive (+) or a negative (–) electrical charge respectively.

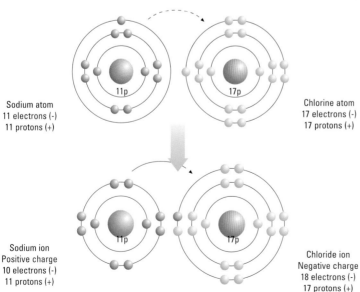

Sodium atom
11 electrons (-)
11 protons (+)

Chlorine atom
17 electrons (-)
17 protons (+)

Sodium ion
Positive charge
10 electrons (-)
11 protons (+)

Chloride ion
Negative charge
18 electrons (-)
17 protons (+)

Structure

Schematic representation of a cell, showing its most important elements

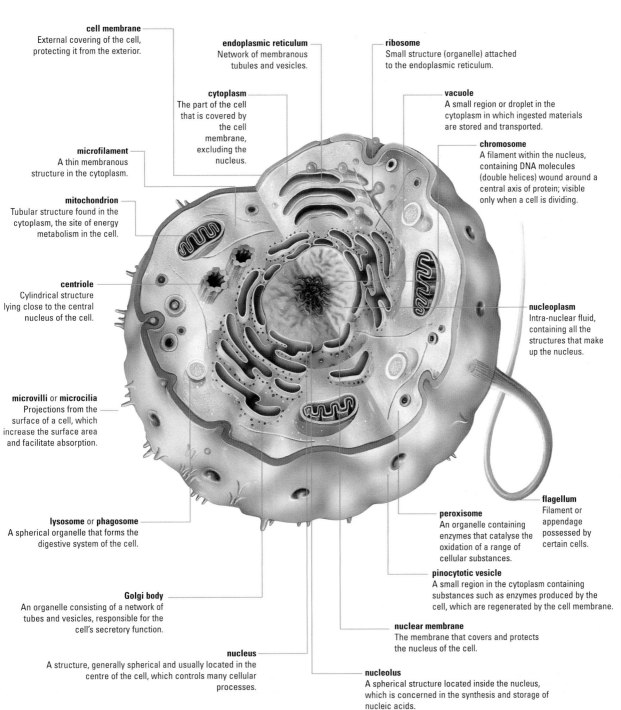

cell membrane
External covering of the cell, protecting it from the exterior.

endoplasmic reticulum
Network of membranous tubules and vesicles.

ribosome
Small structure (organelle) attached to the endoplasmic reticulum.

cytoplasm
The part of the cell that is covered by the cell membrane, excluding the nucleus.

vacuole
A small region or droplet in the cytoplasm in which ingested materials are stored and transported.

microfilament
A thin membranous structure in the cytoplasm.

chromosome
A filament within the nucleus, containing DNA molecules (double helices) wound around a central axis of protein; visible only when a cell is dividing.

mitochondrion
Tubular structure found in the cytoplasm, the site of energy metabolism in the cell.

centriole
Cylindrical structure lying close to the central nucleus of the cell.

nucleoplasm
Intra-nuclear fluid, containing all the structures that make up the nucleus.

microvilli or **microcilia**
Projections from the surface of a cell, which increase the surface area and facilitate absorption.

flagellum
Filament or appendage possessed by certain cells.

lysosome or **phagosome**
A spherical organelle that forms the digestive system of the cell.

peroxisome
An organelle containing enzymes that catalyse the oxidation of a range of cellular substances.

pinocytotic vesicle
A small region in the cytoplasm containing substances such as enzymes produced by the cell, which are regenerated by the cell membrane.

Golgi body
An organelle consisting of a network of tubes and vesicles, responsible for the cell's secretory function.

nuclear membrane
The membrane that covers and protects the nucleus of the cell.

nucleus
A structure, generally spherical and usually located in the centre of the cell, which controls many cellular processes.

nucleolus
A spherical structure located inside the nucleus, which is concerned in the synthesis and storage of nucleic acids.

Cells

The cell

The functions of the cell

The cell is the structural unit of the human organism. It consists of a central nucleus covered by a membrane, lying with various different organelles in an internal fluid, the cytoplasm, which in turn is covered and protected by the cell membrane. The nucleus contains the chromosomes, which play an essential part in cellular reproduction.

A principal function of the cell is to produce energy. In the course of this it consumes oxygen and eliminates carbon dioxide (CO_2). The cell also absorbs carbohydrates, fats, and proteins. The mechanism of uptake may be by diffusion across the membrane, or by the process known as pinocytosis (engulfing). Plant cells, by contrast, consume carbon dioxide in photosynthesis, and in turn release oxygen into the atmosphere.

Another important function of the cell is reproduction or cell division. One mechanism, known as mitosis, is illustrated opposite. It starts inside the nucleus and develops in a sequence of distinct phases.

Pinocytosis

The mechanism by which the cell's nutrients are engulfed

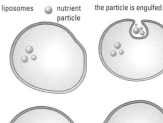

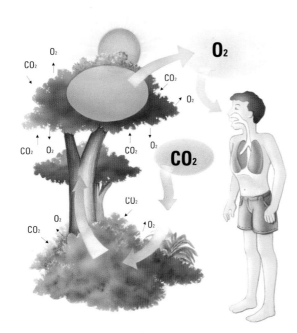

Production of O_2 and consumption of CO_2 by photosynthesizing plants

Reproduction

Cellular reproduction by mitosis

INTERPHASE
The period of time between mitotic divisions. During this period, the cell grows and carries out its normal metabolic functions.

START OF THE PROPHASE
This is the first phase of mitosis and the longest. Pairs of chromatin filaments appear and transform themselves into chromosomes, each consisting of two identical filaments (chromatids).

END OF THE PROPHASE
The nuclear membranes rupture and the cell nucleus disappears as a differentiated structure. The centrioles move to opposite poles of the cell. A colorless "spindle" structure appears inside the cell.

METAPHASE
This is the second phase of mitosis. The chromosomes group themselves around the center of the spindle, or the "equator" of the cell, equidistant from the poles and arrange themselves in a longitudinal pattern that is referred to as the equatorial plate. The centrioles are located at the poles.

ANAPHASE
The paired chromosomes suddenly separate, forming two complete systems, one of which moves towards each centriole. From this moment on, each chromatid becomes a chromosome.

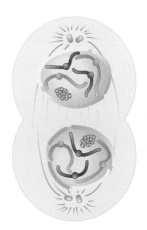

TELOPHASE
Once the chromosomes have repositioned, they again take up the shape of diffuse chromatin filaments. A new nuclear membrane develops to cover each of the new chromosome groups. The nucleoli reappear in the nuclei and the spindle disappears.

Cells

The cell

The tissues

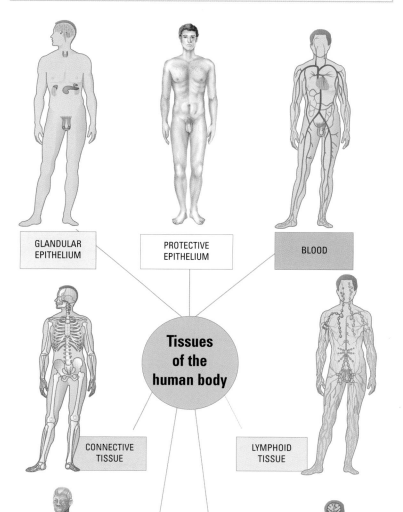

The different types of cell, the structural units or "building blocks" of the human body, together make up the tissues.

Tissues

Tissues of the human body

TYPE	FUNCTION	EXAMPLES
Protective epithelium	protection and insulation from the outside	skin mucous membranes
Glandular epithelium	production of secretions to the outside production of secretions into the blood	exocrine glands endocrine glands
Connective tissue	sustains and supports the covering of organs and reserves of energy and fat	bones cartilage tendons membranes fat
Muscular tissue	mechanical traction for movement	striated muscle smooth muscle
Nervous tissue	transmission of electrochemical stimuli	neurons
Lymphoid tissue	defense against invasive agents from the outside	bone marrow lymph nodes tonsils thymus
Blood	transport of nutrients, hormones, etc.	blood

Cells

The tissues

The tissues

Protective epithelial tissues cover the body and line the organ cavities (skin and mucosa).

Glandular epithelial tissues are specialized for the production of secretions.

Connective tissues provide support and structure (bones, cartilage, tendons). Adipose tissue is a special type of connective tissue; it contains the fats that act as an energy reserve for the entire organism.

Muscular tissue makes up the muscles, which may be either striated or smooth. The heart consists entirely of specialized muscular tissue.

Nervous tissue is composed of cells called neurons, which are specialized for the transmission of nerve pulses.

Lymphoid tissue is located in the bone marrow, the lymph nodes, and the tonsils.

Blood is a tissue consisting of a liquid part, called plasma, which contains a large number of different types of blood cell. Blood is the principal means of transport in the human body.

The various different types of tissues form organs, which are grouped together to form systems with well-defined functions.

Organs and systems

The different tissues make up the organs, which are grouped into systems with well-defined functions. Thus the digestive system, the cardiovascular system, the respiratory system and the blood have a predominantly nutritional function; the urinary system performs the tasks of cleansing and excretion; the skin and the immune system fulfill a basically defensive function; the musculoskeletal system provides for the functions of support and movement; the endocrine system regulates the functions of other systems; the reproductive system is responsible for the perpetuation and reproduction of the species; and finally, the nervous system coordinates the activity of the body as a whole, receiving sensations from the outside by means of the sense organs.

Systems

The reproductive system is responsible for the perpetuation and reproduction of the species; and in the final analysis the nervous system is the ultimate controller and coordinator of the organism as a whole.

The systems of the human body

NUTRITIONAL SYSTEMS	digestive system respiratory system cardiovascular system blood
EXCRETORY SYSTEMS	urinary system
DEFENSIVE SYSTEMS	skin immune system
SUPPORTIVE SYSTEM	musculoskeletal system
REGULATORY SYSTEM	endocrine system
REPRODUCTIVE SYSTEM	reproductive organs
COORDINATION AND CONTROL	nervous system

Skin

The skin

Structure and functions

The skin is the covering that completely envelops the human organism. The component areas and sections can be readily distinguished (see the diagram opposite). It is a genuine organ in its own right, with precise and specific functions. It can also be the site of a large number of symptoms, some specifically skin-related disorders, while others relate to internal problems that involve the skin and are manifested on it. The basic functions of the skin are as follows.

Protection: the skin acts as a defensive element, protecting the body from trauma and preventing the entry of foreign substances and micro-organisms.

Thermoregulation: the skin helps to keep the body's temperature constant (at about 37°C). The initial mechanism for dealing with changes in temperature is increasing or reducing the flow of blood that reaches the skin (referred to as peripheral vasodilation or vasoconstriction). The well-known reddening of the skin in hot weather is the effect of the increased blood circulation; likewise the pallor resulting from chilling is due to the reduced circulation. Sweat also plays a part in regulating temperature; by its evaporation from the skin surface, it removes heat from the organism and brings down the body temperature.

Touch: the skin contains a large number of nerve endings, contained in the tactile corpuscles (see page 25). These detect and pass on to the nervous system the various physical stimuli they encounter, so allowing the brain to receive the sensations of pain, pressure, heat, cold, and so forth, and to take the appropriate action. This makes the instant withdrawal reflex response to such sensations as pricking, burning and so on possible.

Secretion: the skin is provided with a variety of secretion mechanisms. The sweat glands secrete sweat. Sebum, a fatty substance produced from beneath the skin by the sebaceous glands, acts as an insulator and antiseptic, and keeps the skin smooth. Certain skin cells also secrete the pigment melanin, which is responsible for the color of the skin. Finally, the skin cells produce the protein keratin which, by means of a process known as keratinization, hardens and calluses the skin surfaces at the furthest extremities of the body, and makes a major contribution to the replacement of its cells.

The initial mechanism for dealing with changes in temperature is to increase or reduce the flow of blood to the skin, respectively referred to as peripheral vasodilation or vasoconstriction.

Structure

Names of some areas of the skin

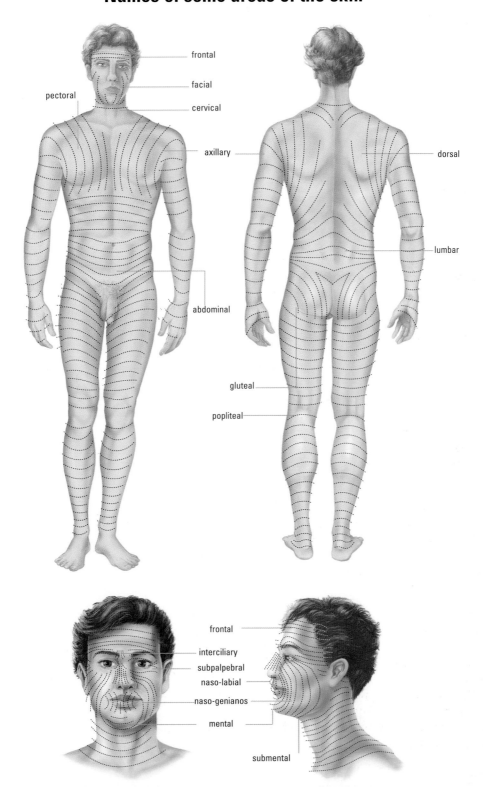

Skin

The skin

Anatomical description

Under the microscope, three superimposed layers can be distinguished in the skin, as follows:

The epidermis is the outermost part of the skin. It is an epithelial layer, about as thick as a sheet of paper, and is made up largely of cells called keratinocytes, which are constantly being renewed. Dispersed between them are the melanocytes, the cells that produce melanin, and the cells of Langerhans, which have a defensive function. The epidermis is made up of four layers or strata: the stratum germinativum, the stratum spinosum, the stratum granulosum and the stratum corneum, each having cells with a different degree of keratinization.

The epidermis does not contain any blood vessels or lymph nodes, but it does contain a large number of nerve endings.

The dermis is the layer below the epidermis. It is rich in blood vessels and has a large number of nerve endings (touch corpuscles), which recognize tactile sensations (Meissner), cold (Kraus), and pressure (Pacini).

This layer also contains the sweat glands, the sebaceous glands, and the hair roots. It is also rich in cells of the immune system, such as mastocytes and histiocytes, structural cells such as fibroblasts and collagen and elastin fibers which are responsible for the skin's elasticity and flexibility.

The hypodermis is the deepest layer of the skin. It is rich in fatty tissue, which acts as a pad protecting the organs beneath it. Below the hypodermis is the subcutaneous tissue, which forms the internal boundary of the skin.

The structure of the skin covering the orifices of the body (the mouth, the nostrils, etc.) is distinctive; it is known as the mucous membrane. The structure of the mucous membrane differs from that of the skin in that, under normal conditions, the mucous membranes have no stratum corneum or stratum granulosum (and thus have a smoother and less hardened appearance than that of skin). Their ancillary organs are distinct from those of the skin and are adapted to the specific functions of the membranes (ancillary salivary glands in the mucous membranes of the mouth, mucous glands in the nose, the taste buds of the tongue and so on).

The structure of the skin covering the orifices of the body (the mouth, the nostrils, etc.) is distinctive; it is known as mucous membrane.

Structure of the skin and the stages of its keratinization

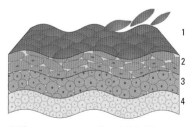

*1. The **stratum corneum**, formed of dead keratinocytes produced in the skin, has a horny consistency. The dead cells are replaced from below by new cells.*

*2. The **stratum granulosum**. Here the keratinocytes stop dividing, lose their nuclei and begin ageing.*

*3. The **stratum spinosum**. The keratinocytes here are in the reproduction stage.*

*4. The **stratum germinativum**, where there is intense multiplication of the keratinocytes.*

Structure

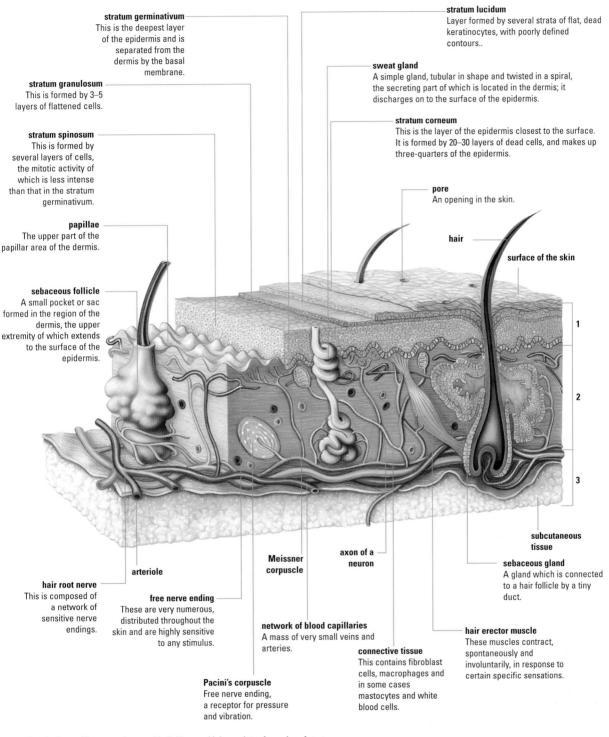

stratum germinativum
This is the deepest layer of the epidermis and is separated from the dermis by the basal membrane.

stratum granulosum
This is formed by 3–5 layers of flattened cells.

stratum spinosum
This is formed by several layers of cells, the mitotic activity of which is less intense than that in the stratum germinativum.

papillae
The upper part of the papillar area of the dermis.

sebaceous follicle
A small pocket or sac formed in the region of the dermis, the upper extremity of which extends to the surface of the epidermis.

stratum lucidum
Layer formed by several strata of flat, dead keratinocytes, with poorly defined contours..

sweat gland
A simple gland, tubular in shape and twisted in a spiral, the secreting part of which is located in the dermis; it discharges on to the surface of the epidermis.

stratum corneum
This is the layer of the epidermis closest to the surface. It is formed by 20–30 layers of dead cells, and makes up three-quarters of the epidermis.

pore
An opening in the skin.

hair

surface of the skin

subcutaneous tissue

sebaceous gland
A gland which is connected to a hair follicle by a tiny duct.

hair erector muscle
These muscles contract, spontaneously and involuntarily, in response to certain specific sensations.

connective tissue
This contains fibroblast cells, macrophages and in some cases mastocytes and white blood cells.

network of blood capillaries
A mass of very small veins and arteries.

Pacini's corpuscle
Free nerve ending, a receptor for pressure and vibration.

axon of a neuron

Meissner corpuscle

free nerve ending
These are very numerous, distributed throughout the skin and are highly sensitive to any stimulus.

arteriole

hair root nerve
This is composed of a network of sensitive nerve endings.

1. **epidermis.** Formed by a very dense epithelial layer, which consists of a series of strata.

2. **dermis.** The second stratum of the skin, formed by a layer of fatty connective tissue, which is both resistant and flexible.

3. **hypodermis.** The lowest part of the dermis, rich in fatty tissue, which provides a protective function for the organs located beneath it.

Skin

The skin

Organs of the skin

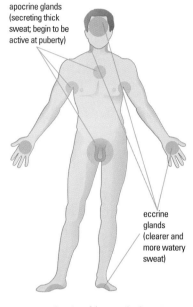

Location of the sweat glands.

Several important appendages are located in the skin, including hairs, nails, sebaceous glands and sweat glands.

Hairs are keratinized structures, more or less cylindrical in shape. Each consists of two parts: the shaft, the part of the hair that projects above the skin surface and the root, which is the part that forms the sebaceous hair follicle, a tubular structure covered by a fibrous sac extending as far as the dermis, where it widens out and forms the hair bulb. Secretions by the sebaceous glands are poured into the interior of the follicle. Attached to each follicle is a hair erector muscle, which raises the hair partly erect in cold conditions or in frightening situations, an action that gives the skin the appearance of "goose pimples."

The hair is formed by a central medulla surrounded by the cortex, a layer rich in melanin (which is responsible for the hair color), which is covered by an external layer or cuticle.

The nails are structures formed by keratinized cells located in the nail base. The part of the nail that is hidden in the base is called the root, and the visible part is referred to as the nail body. At the base of the body is the lunula, a semicircle of a lighter color. The skin base over which the whole nail extends is called the nail bed.

The sebaceous glands are associated with the hair follicles and are more plentiful in areas where the hair grows thickly (the scalp, armpits, genital areas, etc.). They are sac-like cavities lined with cells that produce a fatty secretion, the sebum, which passes to the sebaceous hair follicle and thence to the surface of the skin.

The sweat glands are spiral, tubular structures located in the dermis. There are two types: the apocrine and the eccrine glands. The former secrete a thick sweat, while the latter produce a clearer and more watery form.

Sweat is a fluid consisting mostly of water, but also contains other substances such as ammonia, salts, fats and so on. Sweat is initially odorless, but the actions of certain skin bacteria cause it to decompose, with an accompanying characteristic odor. As has already been pointed out, the secretion of sweat is a means of regulating the body's temperature.

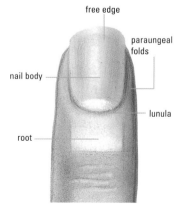

External view of a finger showing the structure of the nail.

Structure

Structure of a hair

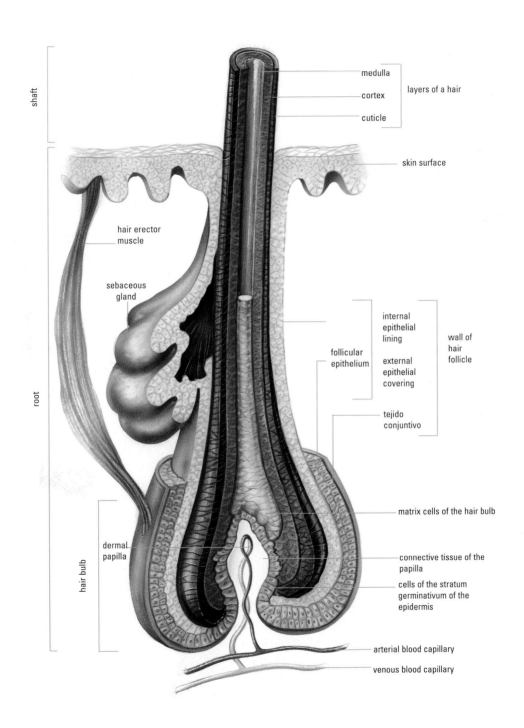

The blood

The blood is a viscous liquid, red in color, which flows through the blood vessels and the heart. Its main task is to carry oxygen and the various other nutrient substances to all the parts of the body.

Blood consists partly of a liquid called the plasma, and partly of blood cells of various kinds: erythrocytes (red blood cells), leukocytes (white blood cells), and thrombocytes (platelets). New blood cells are continuously being regenerated by a process referred to as hematopoiesis, which principally takes place in the bone marrow.

Functions of the blood

The most important functions of the blood are:

Transport of oxygen by the hemoglobin contained in the red blood cells.

Transport of nutrients and hormones dissolved in the plasma. Nutrients include, for example, glucose and vitamins.

Transport of waste products, the result of cellular metabolism, to the excretory organs. Carbon dioxide is excreted via the lungs; for substances excreted in the urine, the excretory organs are the kidneys.

Defense against infection: the leukocytes are responsible for preventing any outside agents that may invade the organism from causing disease. These cells have the ability to identify any substance that does not belong to the body, and set in motion a series of processes that will lead to their destruction or neutralization. Once these cells have accomplished their mission they undergo a process of degeneration, which results in the formation of pus.

Coagulation: the process by which the blood forms a clot. This mechanism takes place by the action of thrombocytes and a series of coagulation agents contained in the plasma. The clot prevents blood escaping from the body, for example through a cut or wound.

Blood consists partly of a liquid called the plasma, and partly of blood cells of various kinds: erythrocytes (red blood cells), leukocytes (white blood cells), and thrombocytes (platelets).

Coagulating blood, in which the clotting process is visible

Functions

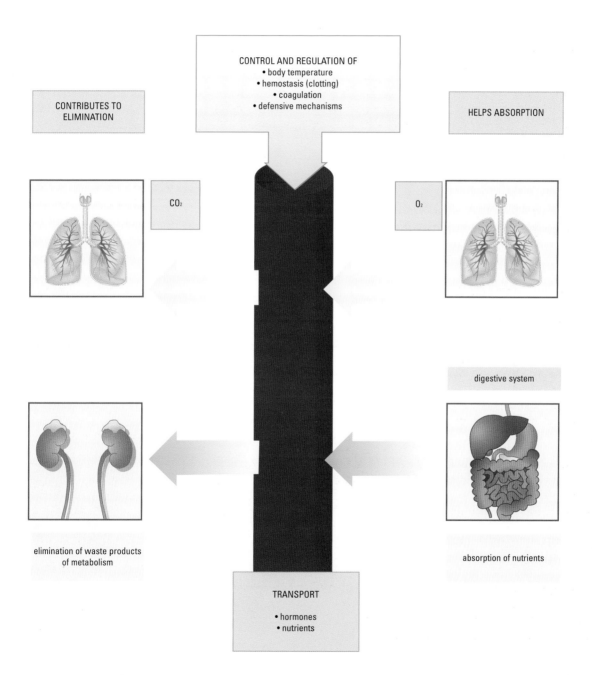

Blood

The blood

Regulation of body temperature: the human body functions under optimum conditions at a temperature of about 37°C. When the ambient temperature is high, the peripheral blood vessels dilate, ensuring a greater flow of blood to the surface areas of the organism, which causes a greater loss of heat. When the ambient temperature is low, the blood vessels contract, delivering less blood to those areas, and the heat loss is reduced.

Composition of the blood

In general terms, blood has two components: plasma and its constituent substances, and the blood cells.

Plasma

The plasma is the liquid part of the blood. If blood is centrifuged, a sediment soon forms, with a yellowish supernatant liquid above it. This liquid is the plasma.

Plasma consists mainly of water containing a large number of dissolved substances, such as ions, glucose, proteins, vitamins, and so on. The proteins contained in the plasma are of three types:

• *Albumin*. This is a very abundant protein. The force of attraction that it exerts on the water, known as osmotic pressure, helps to prevent water from leaving the blood vessels.

• *Globulins*. There are three groups of globulins, known as alpha, beta, and gamma globulins. The first group has a role in substance transport and the others have important functions in the immune system. They constitute the antibodies, substances that are vital to the defense of the organism. The gamma globulins are of various types (A, M, G, E, etc.), each being involved in various immunological processes.

• *Fibrinogen*. This protein is concerned in the mechanisms of blood coagulation. It has first to be activated by a series of substances (activators), which are also contained in plasma.

> If blood is centrifuged, a sediment soon forms, with a yellowish supernatant liquid above it. This liquid is the plasma.

Structure

Components of the blood system

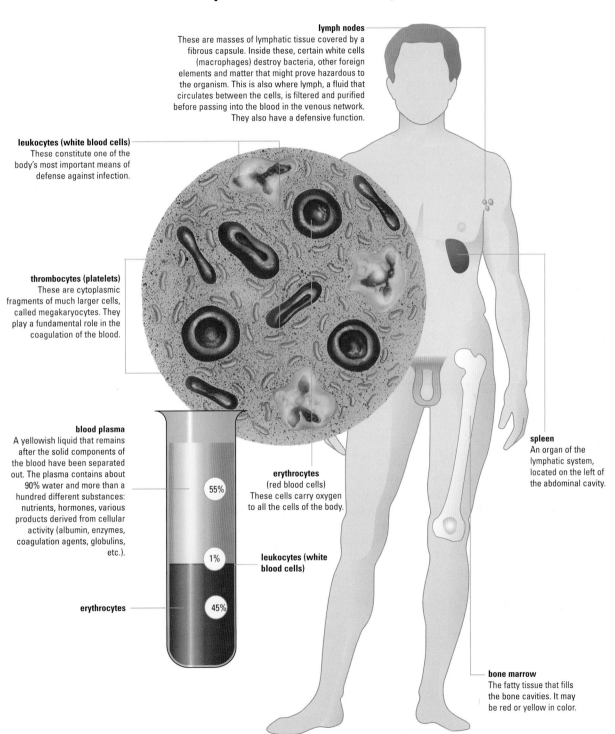

lymph nodes
These are masses of lymphatic tissue covered by a fibrous capsule. Inside these, certain white cells (macrophages) destroy bacteria, other foreign elements and matter that might prove hazardous to the organism. This is also where lymph, a fluid that circulates between the cells, is filtered and purified before passing into the blood in the venous network. They also have a defensive function.

leukocytes (white blood cells)
These constitute one of the body's most important means of defense against infection.

thrombocytes (platelets)
These are cytoplasmic fragments of much larger cells, called megakaryocytes. They play a fundamental role in the coagulation of the blood.

blood plasma
A yellowish liquid that remains after the solid components of the blood have been separated out. The plasma contains about 90% water and more than a hundred different substances: nutrients, hormones, various products derived from cellular activity (albumin, enzymes, coagulation agents, globulins, etc.).

erythrocytes
(red blood cells)
These cells carry oxygen to all the cells of the body.

leukocytes (white blood cells)

erythrocytes

55%
1%
45%

spleen
An organ of the lymphatic system, located on the left of the abdominal cavity.

bone marrow
The fatty tissue that fills the bone cavities. It may be red or yellow in color.

Blood

The blood

COMPONENTS OF THE BLOOD

The blood contains three types of cell: erythrocytes, leukocytes and thrombocytes. These represent about 40 % of the volume of the blood, a percentage known as the hematocrit volume. All these cells are produced in the bone marrow in various parts of the body.

Erythrocytes (red blood cells) have the shape of a biconcave disk. They do not have a nucleus and are about 8 microns in diameter. There are about 4,500,000 to 5,500,000 erythrocytes per cubic millimeter of blood. Erythrocytes contain hemoglobin, which is the molecule responsible for carrying oxygen and which gives the red cells their color. The blood of an adult human contains between 12 and 15 g of hemoglobin per 100 ml.

Leukocytes (white blood cells) are cells with nuclei, and their task is principally defensive. There may be between 4,000 and 11,000 leukocytes per cubic millimeter of blood.

There are three different types of leukocyte: granulocytes, lymphocytes and monocytes.

• Granulocytes are young cells with a single nucleus in the shape of an inverted curve, which may be strip- or rod-shaped or segmented. Segmented leukocytes are cells about 10–14 microns in diameter, with a nucleus divided into several lobes. Depending on the coloration given by the granules of the leukocytes with certain stains, we speak of neutrophils, basophils, and eosinophils, having cytoplasmic granules staining respectively violet, dark purple, and yellowish red.

• Lymphocytes are cells with diameters in the range of approximately 7–20 microns. They are produced in the bone marrow, and then locate to the lymph nodes. There are two types: T lymphocytes, which take part in the cell-mediated immune response, and B lymphocytes, which are responsible for producing antibodies (immunoglobulins).

• Monocytes are the largest of the blood cells, with a diameter of 16–22 microns. They also have a defensive function.

Thrombocytes are not true cells but parts of the cytoplasm of certain very large bone marrow cells, the megakaryocytes. Thrombocytes are only 3 or 4 microns in size, and their main task is to form a plug (clot) when a hemorrhage occurs. There are between 140,000 and 450,000 thrombocytes per cubic millimeter of blood.

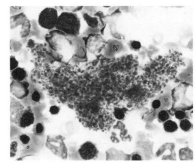

Platelets. These are the smallest elements in the blood, and play a part in the coagulation process.

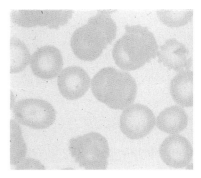

Red cells or erythrocytes.

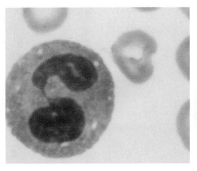

Eosinophil. A blood cell which contains eosinophilic granulations in its protoplasm.

The spleen

The spleen is an organ of the lymphatic system, located in the upper left-hand part of the abdominal cavity. The upper surface is convex in shape, and the lower surface is flat. These surfaces are smooth and adapt to the shape of the organs against which they lie. The blood vessels that supply the spleen are essentially the splenic artery and vein, which pass through the splenic hilus. The spleen of an adult weighs about 200 g.

Spleen tissue, known as the splenic pulp, can be divided into two parts: the red pulp, which contains a large number of blood vessels, and the white pulp, rich in lymphocytes. The great number of blood vessels and the special structure of the red pulp mean that the spleen can serve as a reservoir for blood; it also has an important role in blood regeneration, since it is the organ in which most dead erythrocytes are finally destroyed. The white pulp is the part responsible for carrying out the immune functions of the spleen. This is where the blood is filtered, resulting in the neutralization of foreign substances that may be present in the bloodstream.

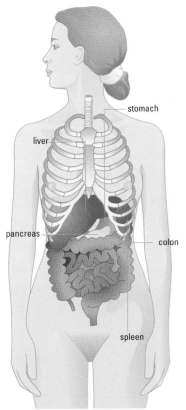

Diagram showing the location of the spleen within the abdominal cavity

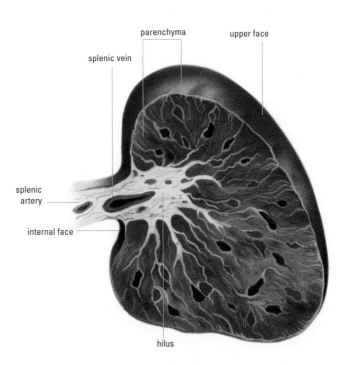

Structure of the spleen

Blood

The blood

Lymph nodes

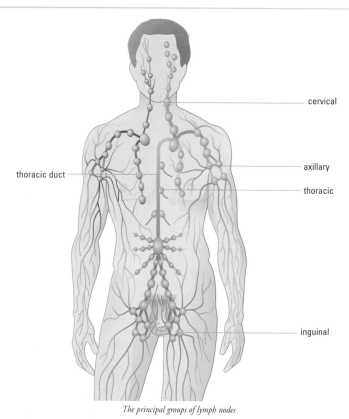

The principal groups of lymph nodes

The lymph nodes are distributed throughout the lymphatic system. They may be superficial or located deep in the tissues, and are grouped into distinct regions: occipital, cervical, submaxillary, axillary, inguinal, etc.

The bone marrow

Inside the bones, between the laminar and trabecular layers that form the bone tissue, there is a spongy substance called the bone marrow. Although this plays a very important part in the formation of bone tissue, the reason why it is included in this chapter is its capacity to produce new blood cells by the process known as hematopoiesis. The bone marrow consists of a fine network of blood vessels extending throughout the marrow cavity, together with a large number of cells, principally hematopoietic cells, distributed throughout this network. These have the capacity to produce erythrocytes, leukocytes, and thrombocytes, and also cells called adipocytes, which contain fat. Adipocytes make up the yellow marrow; they do not have the capacity to create blood cells.

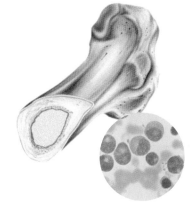

Bone marrow

THE ERYTHROCYTE CYCLE

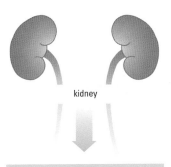
kidney

inactive erythrocytes

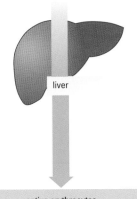
liver

active erythrocytes

stimulation

BONE MARROW
production of erythrocytes

SPLEEN

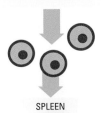

DESTRUCTION

The blood cycle

HEMATOPOIESIS

Blood cells constantly have to be regenerated, as they have only a limited life span and are destroyed at the end of it, mainly in the spleen and the bone marrow. The process of forming new cells takes place in the bone marrow and is called hematopoiesis. The details of the process are different for each type of cell.

The parent cells (stem cells) in the bone marrow are able to provide any of the types of blood cell; they produce whichever type the body needs at a particular moment.

The process by which new erythrocytes are formed is called erythropoiesis, that for forming new leukocytes is known as leukopoiesis, and the term applied to the creation of new thrombocytes is thrombopoiesis.

ERYTHROPOIESIS

The rate at which new erythrocytes are formed in the bone marrow is in relation to the rate at which they are destroyed; this is regulated by means of the action of the hormone erythropoietin, which is secreted by the kidneys. At times when the body has a need for more erythrocytes, production of erythropoietin increases, which stimulates the process of erythropoiesis.

Undifferentiated cells called proerythroblasts are produced by the stem cells. These are large cells with a substantial nucleus, from which erythroblasts are derived which are successively basophilic, polychromatophilic, and acidophilic. Each of these has a smaller nucleus than its predecessor, until the point is reached at which the nucleus disappears. The cells are now called reticulocytes, and are the immediate precursors of the mature erythrocytes, which are discharged into the bloodstream (see page 37). Certain substances play an essential part in erythropoiesis and must be provided by the daily food intake; these include vitamin B_{12}, iron, and folic acid.

Erythrocytes have an average life of 120 days after their entry into the bloodstream and are thereafter destroyed.

Blood

The blood

LEUKOPOIESIS

The undifferentiated stem cells generate cells of considerable size with a central nucleus, known as myeloblasts. These divide to form promyelocytes and then myelocytes, which can adopt three different forms, respectively neutrophilic, basophilic, and eosinophilic. From these arise, through successive phases of maturation, the neutrophil, basophil, and eosinophil metamyelocytes, then the non-segmented granulocytes, and, finally, the forms of parent granulocytes referred to as polynucleocytes (again, neutrophil, basophil, and eosinophil), with their segmented nuclei.

The lymphocytes originate in the bone marrow, but once they are produced they may pass through a phase of maturation in the thymus, and are converted into T lymphocytes. Alternatively, they may mature in the bone marrow itself, and enter the blood in the form of B lymphocytes.

Monocytes are also generated in the bone marrow. The undifferentiated stem cells produce the promonocytes, which subsequently become monocytes and migrate to the tissues. They are also known as macrophages or histiocytes.

THROMBOPOIESIS

Thrombopoiesis, the process by which thrombocytes are regenerated, takes place in the bone marrow, at a rate that is related to the rate at which the platelets already in circulation are being destroyed. It is regulated by the action of a hormone called thrombopoietin, which keeps the rates of destruction and regeneration in balance.

From an initial pluripotent stem cell cells called megakaryoblasts are derived, which give rise in turn to promegakaryoblasts and then megakaryocytes. These finally fragment into thrombocytes, which are released into the bloodstream. Once they are circulating freely, they can be used in the formation of blood clots, should any hemorrhagic lesion or cut occur. Otherwise they pass through a life cycle of about 8 to 10 days and are then destroyed in the spleen.

Blood cycle

General outline of hematopoiesis

The undifferentiated pluripotent stem cell produces, on the one hand, colonies of lymphoid precursors, which give rise to T and B lymphocytes depending on where the cells come to maturity, and, on the other, myeloid precursors from which the proerythroblasts derive (the start of the process of erythrocyte production), the promyelocytes (which give rise to the different types of leukocyte), the megakaryocytes (thrombocytes), and the monoblasts (which finally become the monocytes).

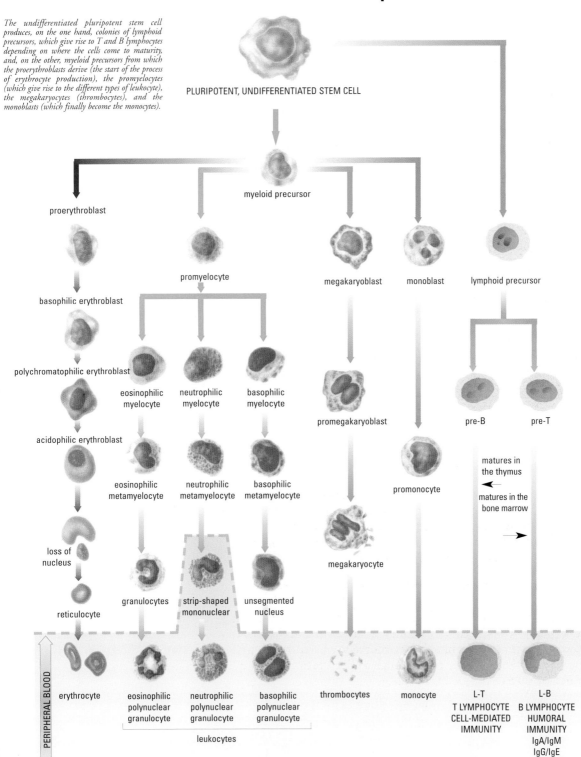

Blood

The blood

Blood groups

During the earliest attempts at transfusing blood from one person to another, it was found that this procedure could give rise to an intense reaction which sometimes ended with the death of the patient. It is now known that these reactions are due to the presence on the erythrocyte surface membrane of a series of substances, known as surface antigens, specific to each person, which act as identifiers for the erythrocytes. Erythrocytes that do not carry the same antigens are regarded as being foreign, and are destroyed by antibodies.

There are many types of surface antigen, which allows for the classification of individuals into different blood groups. The two most important blood group systems, which give rise to the majority of serious transfusion reactions, are the ABO and the Rh groups.

The ABO system. According to this system, the blood of any individual can be classified into one of the following four groups:

Group A. Persons whose erythrocytes carry type A antigens, and lack any other type.

Group B. Persons whose erythrocytes carry type B antigens exclusively.

Group AB. Persons whose erythrocytes carry antigens of both type A and B.

Group O. Persons whose erythrocytes carry neither type A nor type B antigens.

Accordingly, a person can receive erythrocytes from someone else of the same group or of group O, and can donate to a person of the same group or of group AB.

The Rh system. The Rh system is based on the presence or absence in the membrane of the erythrocyte of an antigen referred to as the D antigen. People who have this are classified in the group Rh (D)+, and those without it in the group Rh (D)−.

Compatibility of different blood groups

GROUPS	A	B	AB (UNIVERSAL RECEIVER)	O (UNIVERSAL DONOR)
Antibodies present	Anti-B	Anti-A	None	Anti-A Anti-B
Types of blood that can be received	A O	B O	A, B, AB, O	O

Blood

Blood groups; the ABO system

Identification test for the different groups in the ABO system. If a sample of unknown blood is coagulated by the Anti-A and Anti-B serum, then the sample corresponds to Group A. If coagulation takes place with the Anti-B and Anti-A sera, the blood is Group B. If there are reactions with the Anti-A, Anti-B, and Anti-AB sera, the blood belongs to Group AB. If the blood sample does not react with any of the different sera, it corresponds to Group O.

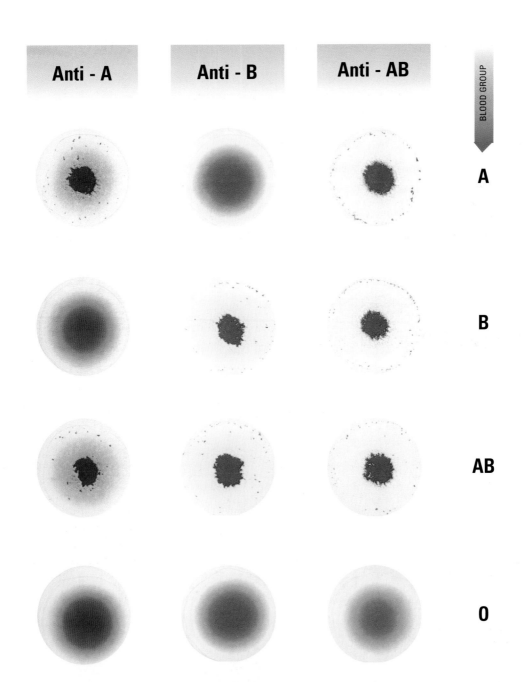

The systems of the human body

The systems

This section is the largest in the book, covering eleven chapters in all, and is the part concerned with the various systems that make up the human organism.

The reader will find here the so-called "traditional" systems that we have all studied in our schooldays, such as the respiratory and digestive systems. We shall also, however, be looking at other systems of which the detailed study may be considered more complex because their component organs and structures have no spatial continuity. These are the immune system and the endocrine system.

The information provided in each chapter makes reference to the anatomical structure and, when required, to the normal functioning of the system.

This book does not claim to provide an exhaustive explanation of the systems in the way that a book for specialists might do. On the contrary, the intention is to provide the general public with some basic knowledge which will allow for the study of the human body in an objective and clear manner, emphasizing those characteristics that are considered essential.

The systems of the human body

The digestive system

Anatomical description

The digestive system consists of the following parts: the buccal (mouth) cavity, the pharynx, the esophagus, the stomach, the small intestine, the large intestine, and the anus. Each of these is a multi-layered structure (see illustration). The small intestine is composed of three parts: the duodenum, the jejunum, and the ileum; the large intestine is also made up of several sections, the cecum, the ascending colon, the transverse colon, the descending colon, and the rectum. Included with these are the glands associated with the duodenum: those of the liver and pancreas, which produce the secretions necessary for the complete digestion of food.

**One of the principal activities of the human body is feeding itself, an essential action for the other actions of the organism to be carried out.
The digestive system consists of the mouth cavity, the pharynx, the oesophagus, the stomach, the small intestine, and the large intestine.**

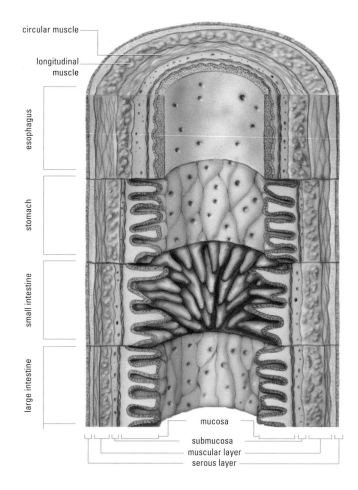

Basic structures of the different sections of the wall of the digestive tract

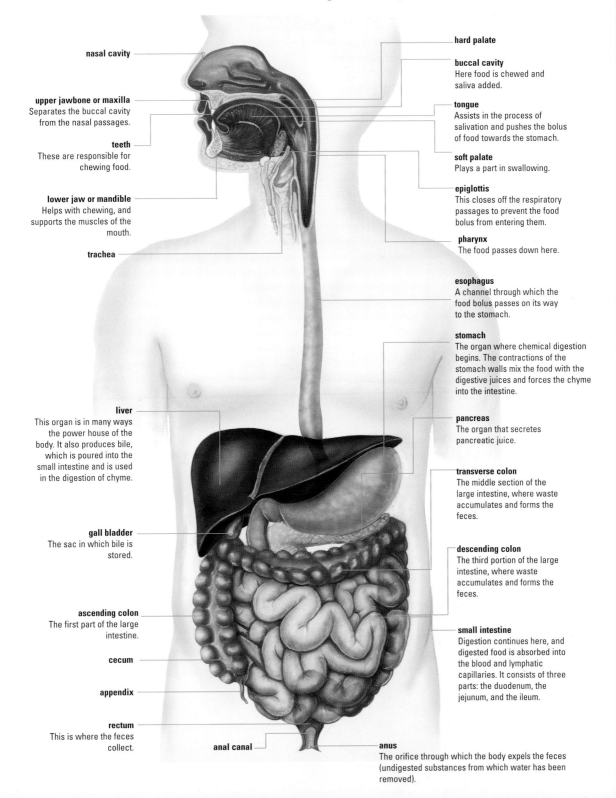

Anatomy

General view of the digestive system

- **nasal cavity**
- **upper jawbone or maxilla** — Separates the buccal cavity from the nasal passages.
- **teeth** — These are responsible for chewing food.
- **lower jaw or mandible** — Helps with chewing, and supports the muscles of the mouth.
- **trachea**
- **liver** — This organ is in many ways the power house of the body. It also produces bile, which is poured into the small intestine and is used in the digestion of chyme.
- **gall bladder** — The sac in which bile is stored.
- **ascending colon** — The first part of the large intestine.
- **cecum**
- **appendix**
- **rectum** — This is where the feces collect.
- **anal canal**
- **hard palate**
- **buccal cavity** — Here food is chewed and saliva added.
- **tongue** — Assists in the process of salivation and pushes the bolus of food towards the stomach.
- **soft palate** — Plays a part in swallowing.
- **epiglottis** — This closes off the respiratory passages to prevent the food bolus from entering them.
- **pharynx** — The food passes down here.
- **esophagus** — A channel through which the food bolus passes on its way to the stomach.
- **stomach** — The organ where chemical digestion begins. The contractions of the stomach walls mix the food with the digestive juices and forces the chyme into the intestine.
- **pancreas** — The organ that secretes pancreatic juice.
- **transverse colon** — The middle section of the large intestine, where waste accumulates and forms the feces.
- **descending colon** — The third portion of the large intestine, where waste accumulates and forms the feces.
- **small intestine** — Digestion continues here, and digested food is absorbed into the blood and lymphatic capillaries. It consists of three parts: the duodenum, the jejunum, and the ileum.
- **anus** — The orifice through which the body expels the feces (undigested substances from which water has been removed).

The systems of the human body
The digestive system

The buccal cavity

The buccal cavity opens to the outside through the mouth (the oral fissure). The palate forms the roof of the mouth, and separates it from the nasal passages. The front two-thirds of the palate consists of bone (the hard palate) and the remainder of membranous muscle (the soft palate). The floor of the cavity is made up of several different muscular bodies anchored to the lower jaw or mandible, above which lies the tongue, and the side walls are formed by the cheeks. At the back the mouth is in contact with the pharynx, via what is referred to as the isthmus of the fauces, which is the extension downwards of the soft palate, the uvula, and the tonsils.

The buccal (mouth) cavity forms the beginning of the digestive system, through which food is introduced to the body.

Anatomy of the buccal cavity

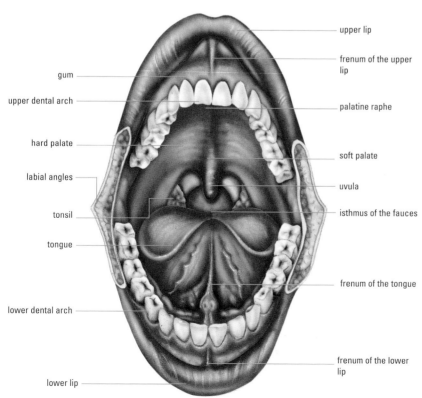

The buccal mucosa

The inside of the buccal cavity is lined with a thin mucous membrane known by the general term of oral mucosa. There are different parts of this mucosa, depending on the specific area covered: the labial mucosa, the lingual mucosa, the palatine mucosa, the gingival mucosa, and the tonsillar mucosa.

The oral fissure

The oral fissure, or the mouth, is an opening in the lower part of the face, surrounded by the upper and lower lip, which come in contact at the labial angles. Various muscles take part in the great mobility and demonstration of expression that can be achieved by the lower part of the face.

Muscles of the mouth

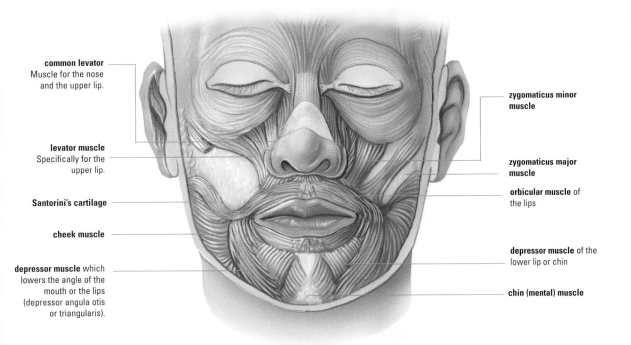

The systems of the human body
The digestive system

Teeth and gums

The gums are semicircular bony structures covered with mucosa, in which the teeth are implanted in cavities referred to as alveoli or sockets. The first teeth (20 in number) break through at the age of about six months, and from the age of six years are progressively replaced by the adult teeth, of which there are 32.

The teeth, and the ages at which they erupt

INCISOR

CANINE

PREMOLAR

MOLAR

PRIMARY DENTITION
(the milk teeth)

INCISORS
central (1) (6–8 months))
lateral (2) (8–10 months))

CANINES
(3) (16–20 months)

MOLARS
first (4) (10–15 months))
second (5) (20–24 months)

PERMANENT DENTITION

INCISORS
central (1) (6–8 years))
lateral (2) (7–9 years))

CANINES
(3) (11–13 years)

PREMOLARS
first (4) (9–11 years))
second (5) (10–12 years)

MOLARS
first (6) (6–7 years)
second (7) (12–17 years)
third, or wisdom teeth (8) (18–30 years; in some people they never appear)

The adult teeth are classified according to their function, as 8 incisors, 4 canines, 8 premolars, and 12 molars. The incisors and the canines tear and bite the food; the premolars and molars grind it up and chew it.

Each tooth consists of three parts: the root, implanted in the socket; the neck; and the crown, the outer part. The shape of each part is distinctive, depending on the type of tooth concerned.

The outer layer of the crown, the enamel, consists of an extraordinarily hard whitish material. The outer layer of the roots of the tooth, which anchor it in the tooth socket, is known as cementum. The intermediate layer is called dentine. The innermost region, tooth pulp, consists of connective tissue in which the blood vessels and nerve endings that make the tooth into a living organ are located.

Structure

Longitudinal section through an incisor tooth

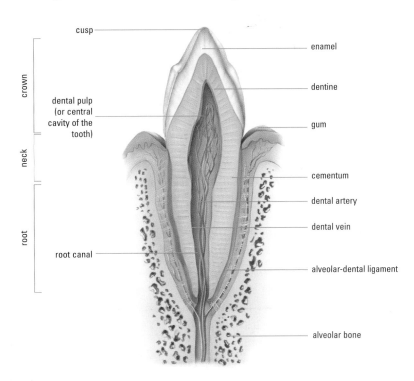

Diagrammatic representation of the longitudinal section of a molar

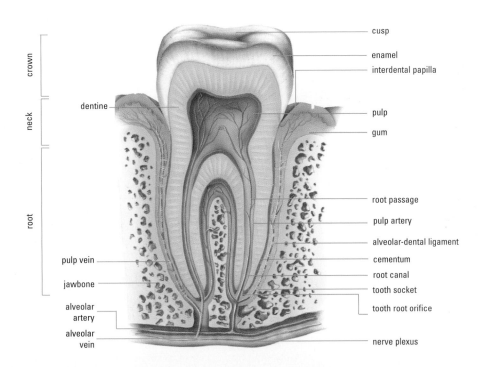

The systems of the human body

The digestive system

The tongue

Supported on the base of the mouth is the tongue, which is a flattish muscular structure. Its muscles are secured at the back to a bone called the hyoid bone, and are free at the front; this structure allows for an extraordinary degree of mobility and enables the tongue to play a major part in swallowing and in phonation. The tongue is also involved in the sense of taste, which is why the lingual mucosa is pitted with a large number of protuberances called taste papillae, which are of several shapes: filiform, fungiform, and calceiform. Each detects a different taste: salty, sour, bitter, and so on. Sweet tastes are sensed at the tip of the tongue, sour tastes in the lateral part, and bitter tastes at the back.

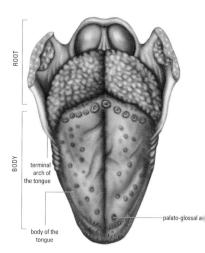

Dorsal view of the tongue

The tongue and palate

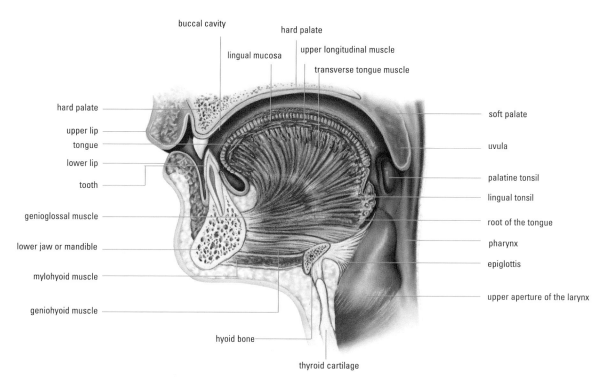

The salivary glands are clusters of structures that secrete saliva, which is essential for the correct functioning of the digestive system.

Tonsils

These are two lymphatic organs located one on each side of the isthmus of the fauces, which may be more or less developed in each individual. They are referred to as palatine tonsils to distinguish them from others in the pharynx. Being lymphoid organs, they have an essentially defensive function.

Pharynx

This is a fibromuscular passage that in the digestive process allows food to pass along it, and, in breathing, allows air to pass in and out.

Salivary glands

There are six of these, and their task is to secrete saliva, which is essential for the correct functioning of the digestive system.

The parotid glands lie at cheek level, and introduce the saliva into the mouth along the Stenon duct.

The submaxillary glands, at the side of the base of the mouth, secrete via the Warton duct.

The sublingual glands are located in the anterior central part of the base of the mouth, and secrete via the Bartholin duct.

A considerable number of glands are spread along the length of the oral mucosa, which mainly have the function of keeping the membrane moist.

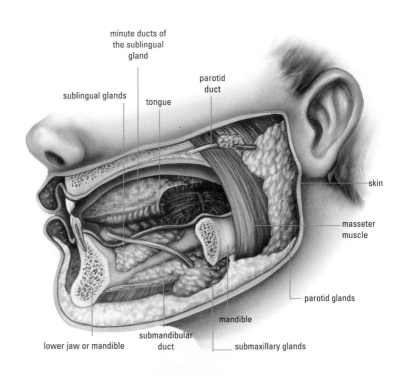

The systems of the human body
The digestive system

The esophagus

The esophagus is a tube about 25 cm long, extending downwards through the thoracic cavity from the pharynx to the stomach. From the neck, the esophagus passes down through the thorax into the chest, in the posterior mediastinic area immediately in front of the vertebral column, crosses the diaphragm muscle that separates the thorax from the abdomen, and, inside this cavity, joins up with the stomach via an orifice called the cardia. It can therefore be said that there are three parts of the esophagus: the cervical, thoracic (the longest), and the abdominal.

Entry to the esophagus is regulated by a ring of muscle, the upper esophageal sphincter, which controls voluntary swallowing. The cardia, the point at which the esophagus enters the stomach, is also sphincter-like, and is usually closed to prevent the contents of the stomach from entering the esophagus.

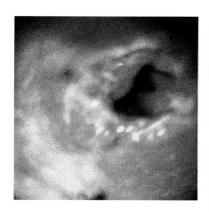

Endoscopic view of the esophagus.

Schematic section through the wall of the esophagus

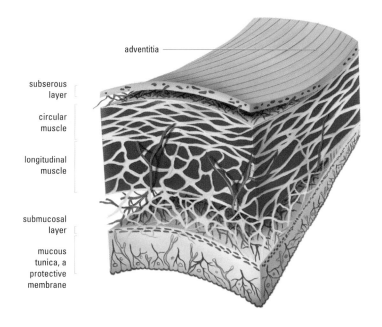

The wall of the esophagus consists of four layers, which from the outside to the inside are respectively an external layer of connective tissue (the adventitia or serosa), an intermediate muscular layer, a submucosal layer containing a large number of secreting glands, and a mucosal layer that lines the whole structure, which when at rest is characterized by many longitudinal folds.

Structure

Views of the esophagus

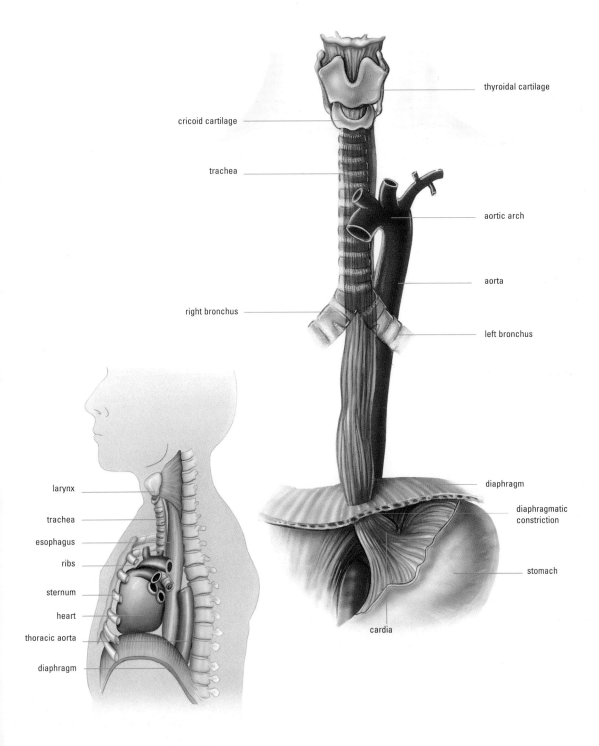

The systems of the human body
The digestive system

The stomach

The stomach forms a sizeable bag into which the esophagus opens via the cardia. Here food is stored temporarily, while the process of digestion begins; the food then leaves the stomach and enters the intestine via the pyloric sphincter. The stomach lies in the upper left-central part of abdominal cavity, below the diaphragm. There are three regions: an upper region called the fundus of the stomach, into which the cardia opens; a central part, the body of the stomach; and a horizontal part that opens to the pyloric sphincter.

Seen from the front, the stomach has a concave exterior on the right side, which runs from the cardia to the pyloric sphincter, and is known as the lesser curvature. On the left it has a convex exterior, which also extends from the cardia to the pyloric sphincter on this side and which is characterized by a considerable curve, called the greater curvature.

The structure of the stomach walls are similar to those of the esophagus, with a serous external layer, a muscular layer, a submucosal layer, and a mucosal layer, in which glands that secrete pepsinogen, hydrochloric acid, water, bicarbonate ions, and mucus are located.

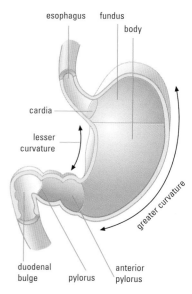

The parts of the stomach

Schematic section through the wall of the stomach

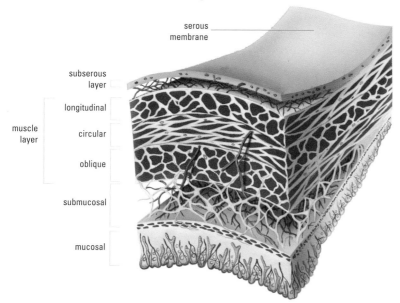

Structure

Structure of the stomach

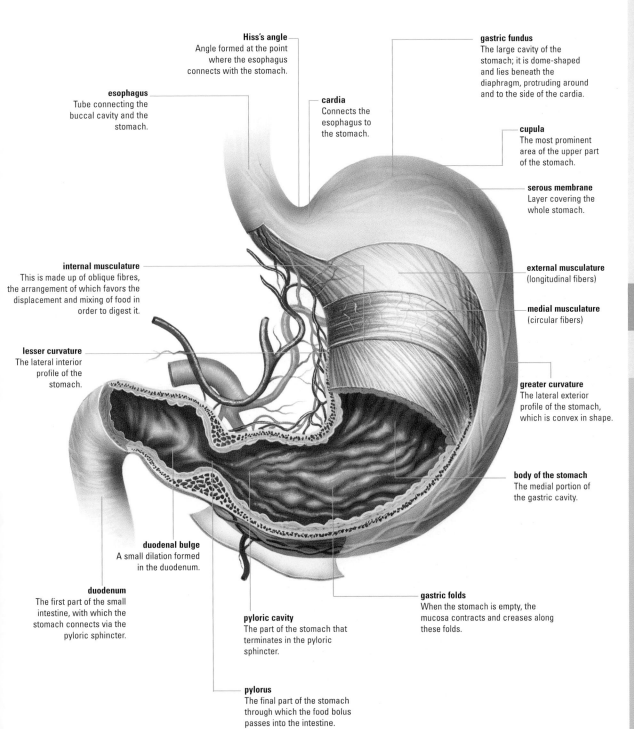

The systems of the human body
The digestive system

The small intestine

This starts from the pyloric sphincter and is about seven meters in length, extending as far as the ileocecal valve. It is divided into three parts: the duodenum, jejunum, and ileum. The duodenum consists of four sections: the first is called the duodenal bulge. The second part lies vertically and the third part horizontally, while the fourth part is connected to the jejunum at what is called the duodeno-jejunal angle or the Treitz angle. Located inside the second part is Vater's ampulla, into which the ducts open which lead from the liver and the pancreas, known respectively as the choledocus or bile duct, and the canal (or duct) of Wirsung.

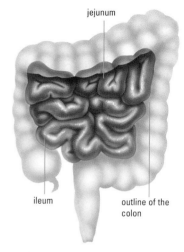

The small intestine: the jejunum and ileum, leading up to the large intestine.

Section through the wall of the small intestine

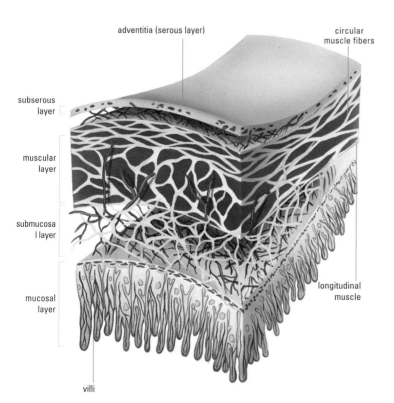

The jejunum and the ileum: from the duodeno-jejunal angle, the small intestine continues in the form of the jejunum and the ileum, which repeatedly form a number of loops. The ileum terminates at the ileocecal valve. The intestinal mucosa is characterized by the presence of valves and also by the intestinal villi, which increase the surface area available for absorption.

Structure

Interior view of the walls of the intestine

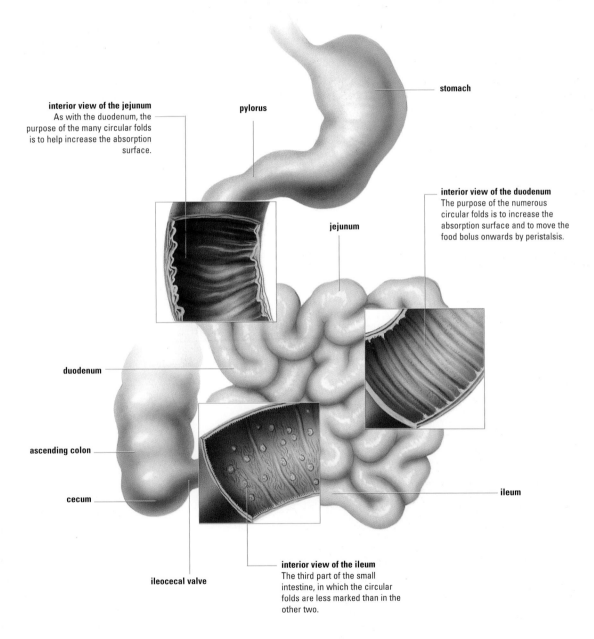

The systems of the human body
Digestive system

The large intestine

This is the final part of the digestive system, starting at the ileocecal valve and ending at the anus. It is subdivided into the cecum, the ascending colon, the transverse colon, the descending colon, the rectum, and the anus.

The cecum contains the vermiform appendix, the function of which is uncertain.

The colon consists of an ascending part, a transverse part, and a descending part, which passes through the sigmoidal colon and ends in the rectum or rectal bulb. The colon is characterized by transverse folds known as haustra.

The rectum is the final section, and opens to the exterior by way of the anus.

The anus is a sphincter that controls the expulsion or retention of the fecal matter.

Endoscopic view of the colon.

Section through the wall of the large intestine

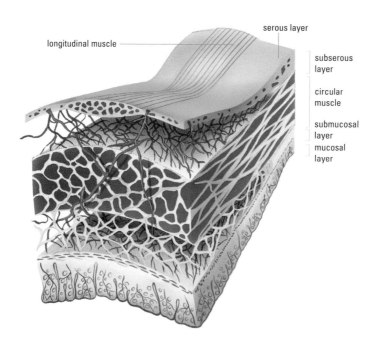

The walls of the large intestine have a serous peritoneal coating, a muscular layer with longitudinal and circular fibres, a submucosal layer, and a mucosal layer rich in tubular mucus-secreting glands.

Structure

The large intestine and its various parts

hepatic flexure or angle
The point where the ascending colon connects with the transverse colon.

transverse colon
Portion of the colon that runs from the right to the left of the abdominal cavity, to the splenic flexure, where it changes direction abruptly, forming a right angle.

splenic flexure
The point where the transverse colon connects with the descending colon. It is located around the spleen.

tinea libera
Line of banded muscle running along the whole of the colon from the cecum as far as the rectum.

haustra
The pouches that make up the large intestine.

descending colon
The part of the colon that forms the continuation from the transverse colon.

ascending colon

paracolic lines
Transverse lines which feature periodically along the entire length of the large intestine, giving it the appearance of a chain of pouch-shaped formations.

ileocecal valve
An orifice connecting the ileum and the cecum; the region where the small intestine connects with the large intestine.

sigmoidal colon
Part of the final section of the descending colon, before it opens into the rectum.

cecum
The initial part of the large intestine (which in this section appears very dilated). The ileocecal valve opens into it.

rectum
The final portion of the digestive tract, which enters the pelvic cavity before opening to the outside through the anus. This is where the fecal matter, the residues of the whole digestive process, is stored until it is expelled.

vermiform appendix
A small cylindrical formation, some 9–10 cm long, lying at the extreme end of the cecum and forming a closed sac. Its function is unknown.

anus
The last point in the digestive tract, consisting of a sphincter structure with a channel some 3 cm long, which by opening and closing regulates the expulsion or retention of the fecal matter.

The systems of the human body

The digestive system

The peritoneum

The peritoneum is a serous membrane that covers the interior of the whole abdominal cavity and many of the organs located in it. It consists of two layers: the external or parietal layer is attached to the walls of the abdominal cavity, while the internal or visceral layer enters this cavity as an extension of the parietal layer and covers the intra-abdominal organs, forming the ligaments between them such as the epiploons, the mesentery, and so on. The parietal layer encloses a space referred to as the peritoneal cavity; normally this contains hardly any liquid, but in certain pathological conditions it may fill up with fluid.

Transverse section through the abdomen; view from above

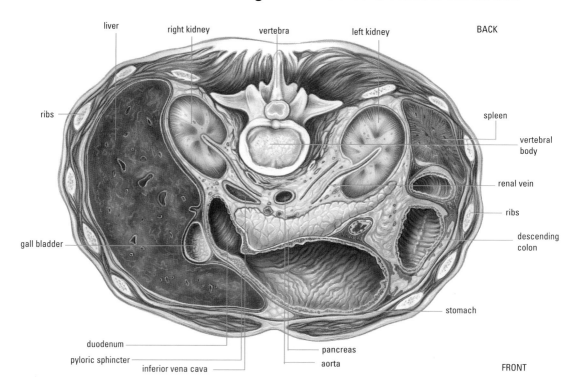

Structure

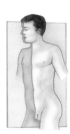

Schematic section through the peritoneal cavity: the peritoneal wall (blue) covers all the intra-abdominal structures

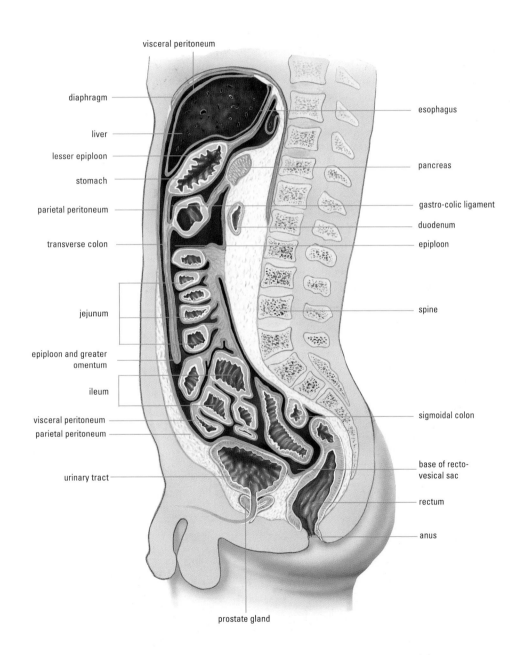

The systems of the human body
The digestive system

The liver

The liver is the largest organ in the human body. It lies in the upper right part of the abdominal cavity, in the area corresponding to the right hypochondrium, and also occupies part of the epigastrium, a region in front of the stomach. Its outer surface has a granular appearance and is dark red in color. The front part is divided into two segments which are clearly separated by the suspensory ligament, which connects it to the diaphragm. These segments are referred to as the left lobe and the right lobe, the latter being much larger than the former. The upper anterior surface is convex in form, and is in contact with the lower surface of the diaphragm and the interior part of the abdominal wall. The posterior surface, which is also convex in shape, is supported by the posterior edge of the diaphragm, to which it is connected by a strong ligament, the coronary ligament, above the inferior vena cava; the latter is joined by the suprahepatic veins, which recover the blood after it has undergone hepatic metabolism. The lower surface is flat and is supported on part of the transverse colon, the apex of the right kidney, and the stomach (to which it is connected by means of the hepato-colic, hepato-renal, and gastro-hepatic ligaments or the lesser epiploon respectively), as well as on the bile duct. In the center of this lower face is the hepatic hilum, the point at which the portal vein and the hepatic artery enter the liver, which carry blood from the spleen, the digestive tract, and the aorta, and through which the bile duct emerges, carrying the bile produced in the liver to the small intestine.

The liver is almost completely covered by the abdominal peritoneum, and is contained entirely by a capsule of fibrous tissue called Glisson's capsule.

Hepatic tissue is made up of small hexagonal lobules, formed by groups of hepatic cells (hepatocytes) around a midlobular vein. Between the hepatic lobules are the perilobular spaces or portal spaces, which contain a branch of the portal vein, another branch of the hepatic artery, and a bile duct. The blood reaches each lobule along branches of the portal vein and the hepatic artery; this blood is filtered between the hepatocytes and passes to the midlobular vein and hence to the suprahepatic veins, which in turn pass the blood (which has already been used in liver metabolism) on to the inferior vena cava. In their turn, the hepatocytes, with products removed from the blood, secrete the bile and discharge it into the bile ducts.

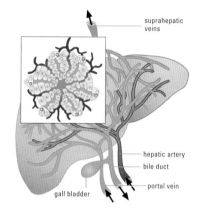

Schematic representation of the internal structure of the liver; hepatic lobe.

The hepatic cells (hepatocytes) secrete bile, which helps with the digestion of fats.

Structure

Front view of the liver

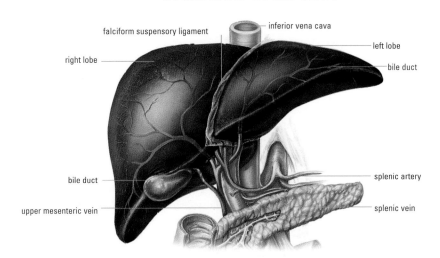

The liver seen from below

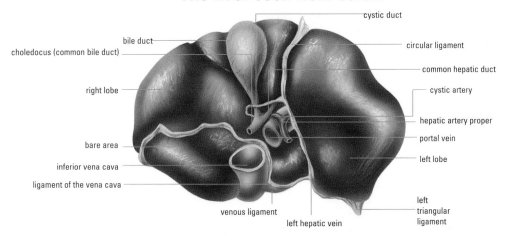

The liver seen from above

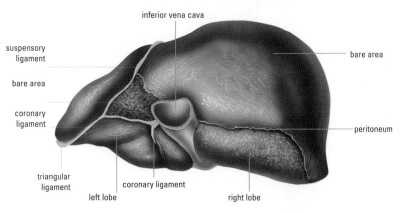

The systems of the human body
The digestive system

The bile ducts

The liver manufactures bile, which is essential for the complete digestion of food. A complex network of channels, the bile ducts, link the points of its secretion – the hepatocytes – to the intestine. The small bile ducts merge to form progressively larger channels which eventually join the right and left main bile ducts, which drain into the common hepatic duct in the hepatic hilum. The hepatic duct is connected to the cystic duct, which leads to the gall bladder. This constitutes a reservoir where the bile is stored and concentrated before being passed on to the intestine. Arising from the junction of the hepatic duct with the cystic duct is the choledocus, which connects to the duct of Wirsung; this runs from the pancreas and opens into the duodenum via Vater's ampulla. At this level, this is called the sphincter of Oddi.

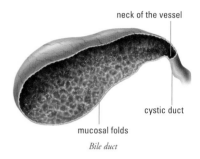
Bile duct

Interior view of the bile ducts

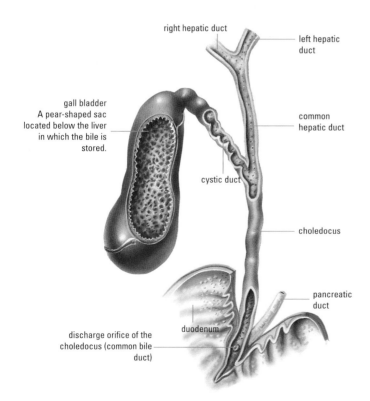

The pancreas

This is a mixed gland: on the one hand, it contains exocrine tissue which produces pancreatic juice, and, on the other, it also contains an endocrine portion formed by the islets of Langerhans, which produce the hormones insulin and glucagon.

The pancreas is an elongated organ consisting of three parts: the head, the body, and the tail (see illustration opposite). The pancreatic juice drains via the confluent ducts to the Wirsung and Santorini collection ducts, which empty into the duodenum (the former at Vater's ampulla and the latter above it).

The pancreas

View of the pancreas

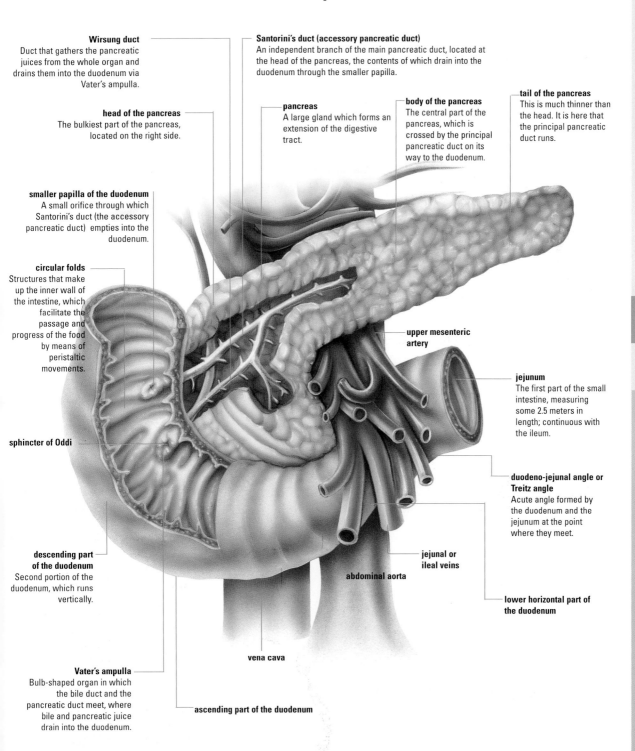

Wirsung duct
Duct that gathers the pancreatic juices from the whole organ and drains them into the duodenum via Vater's ampulla.

Santorini's duct (accessory pancreatic duct)
An independent branch of the main pancreatic duct, located at the head of the pancreas, the contents of which drain into the duodenum through the smaller papilla.

head of the pancreas
The bulkiest part of the pancreas, located on the right side.

pancreas
A large gland which forms an extension of the digestive tract.

body of the pancreas
The central part of the pancreas, which is crossed by the principal pancreatic duct on its way to the duodenum.

tail of the pancreas
This is much thinner than the head. It is here that the principal pancreatic duct runs.

smaller papilla of the duodenum
A small orifice through which Santorini's duct (the accessory pancreatic duct) empties into the duodenum.

circular folds
Structures that make up the inner wall of the intestine, which facilitate the passage and progress of the food by means of peristaltic movements.

upper mesenteric artery

sphincter of Oddi

jejunum
The first part of the small intestine, measuring some 2.5 meters in length; continuous with the ileum.

duodeno-jejunal angle or Treitz angle
Acute angle formed by the duodenum and the jejunum at the point where they meet.

descending part of the duodenum
Second portion of the duodenum, which runs vertically.

jejunal or ileal veins

abdominal aorta

lower horizontal part of the duodenum

Vater's ampulla
Bulb-shaped organ in which the bile duct and the pancreatic duct meet, where bile and pancreatic juice drain into the duodenum.

vena cava

ascending part of the duodenum

The systems of the human body

The digestive system

Function of the digestive system

The main function of the digestive system is to provide the body with nutrition in the form of the nutrients essential for its maintenance and growth. Most food components cannot be used by the body in their natural state, and must be modified inside the digestive tract in order to be utilized as energy. This process consists of various phases: salivation and chewing, which take place in the mouth, and the digestion of the food, which takes place in the stomach and the small intestine. A necessary part of the process is the conveyance of the food through the digestive tract.

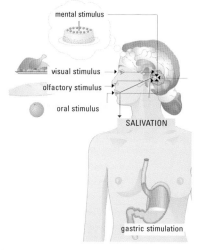

Mechanisms of salivation

Functions of the digestive tract

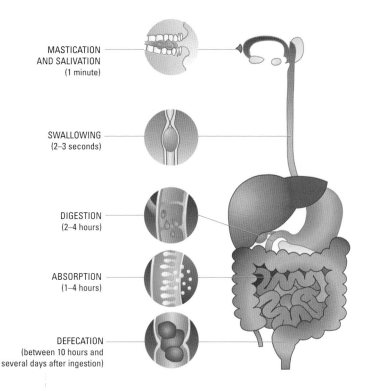

The absorption of nutrients also takes place in the small intestine. Feces are formed in the large intestine, and defecation, or the expulsion of the fecal matter, takes place at the anus.

All these events, with the exception of chewing or mastication, swallowing, and defecation, are under the control of the autonomous nervous system, and are therefore involuntary.

Functions

Swallowing

The passage of the food bolus from the mouth to the esophagus

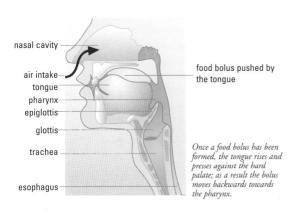

Once a food bolus has been formed, the tongue rises and presses against the hard palate; as a result the bolus moves backwards towards the pharynx.

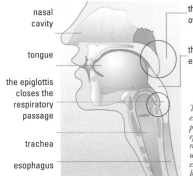

To prevent the food bolus entering the respiratory passages, the uvula rises, the epiglottis closes off the respiratory tract, and the upper sphincter of the esophagus opens to allow the bolus to enter.

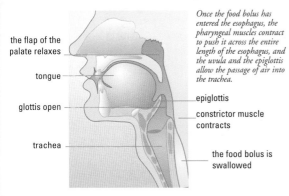

Once the food bolus has entered the esophagus, the pharyngeal muscles contract to push it across the entire length of the esophagus, and the uvula and the epiglottis allow the passage of air into the trachea.

Movements of the stomach

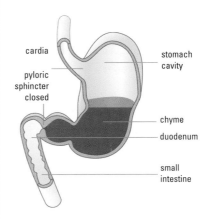

Once the food bolus enters the stomach, it is mixed with the gastric juice secreted by the various gastric glands. The mixture is agitated by the undulating movements of the stomach walls until it is transformed into a dense creamy mixture known as chyme. The pyloric sphincter stays closed in order to prevent the chyme from passing into the intestine.

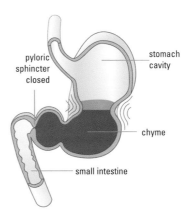

In the lower half of the stomach, the undulating movements (peristalsis) are much more vigorous, which causes the chyme to move towards the pyloric region, the sphincter of which remains closed and prevents the chyme from passing into the intestine.

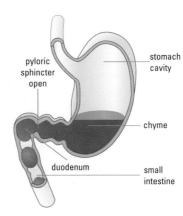

The lower end of the stomach (the pylorus) acts as a valve which, stimulated by the chyme, opens to allow small quantities to pass into the duodenum; further pulsations of the stomach cause most of the contents to pass into the intestine.

The systems of the human body

Nutrition

Nutrition and metabolism

The term nutrition is used to describe all the processes that provide the energy for an organism in order for it to carry out its vital functions. Food supplies the nutrients; the digestive system then prepares them for use by the organism. The term metabolism comprises the whole series of transformations that the food undergoes in order for the body to exploit it. There are certain substances, classified as essential, which the human body is unable to synthesize, and which must be absorbed by way of food. They include certain amino acids, fatty acids, inorganic elements, and vitamins.

Basic nutrients

The basic nutrients are the fundamental substances provided by food which are absolutely necessary for the organism to sustain life. There are three main classes: carbohydrates, including the sugars (the primary sources of energy), proteins (used in tissue formation), and lipids, which include the fats (energy reserves).

Carbohydrates: digestion reduces the polysaccharides to simpler molecules, the monosaccharides, of which the most important is glucose. Glucose is stored in the liver in the form of glycogen; when the body's cells need glucose the glycogen is converted back into glucose by the action of insulin, one of the hormones secreted by the pancreas. The metabolism of the carbohydrates releases energy. In situations where they are not available to the body, the metabolism of fats and proteins provides alternative means of providing energy.

Proteins provide the material that enables cells and tissues to grow and regenerate. Protein molecules consist of chains of peptides, which the processes of digestion transforms into amino acids; these can be absorbed by the body and used in the liver to prepare specific proteins. Some of these amino acids, called the essential amino acids, cannot be synthesized and can only be obtained from food. In addition to their formative and restorative function, proteins can also be used as an

Function

Carbohydrates
Production of energy

Proteins
Formation of new tissues
Replacement of lost material
Alternative source of energy

 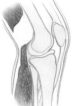

Fats
Formation of certain cellular structures
Formation of hormones
Alternative energy reserve
Intestinal absorption of fat-soluble vitamins
Padding and protection of internal organs

Proteins

Fats

Carbohydrates

alternative energy source when the body does not have carbohydrates available.

Lipids are constituents of the cell membranes, play a part in the synthesis of certain hormones, and are a source of energy. They are major constituents of adipose tissue, which acts as protection for various organs, and are involved in the absorption in the intestine of certain vitamins.

Cholesterol

One special lipid is cholesterol. It circulates in the blood, in association with transporting molecules called lipoproteins. There are two varieties of these: those of low and very low density (known as LDL and VLDL respectively), and those of high density (HDL). Generally, it is the LDL that are responsible for delivering the cholesterol to cells that need it. When the cells cannot absorb any more cholesterol, the quantity of lipoprotein bound to cholesterol in the blood is increased; this accumulates in the arteries, and eventually cause problems with arteriosclerosis. The HDL have the function of delivering the cholesterol to the liver, in order for it to be eliminated in the bile. The cholesterol contained in food is found in fats of animal origin.

Triglycerides are contained in foodstuffs; they are made up of fatty acids chemically combined with glycerol. The fatty acids, depending on their chemical structure, may be classified as saturated or unsaturated fatty acids. The saturated fatty acids are mainly found in animal fats, while the unsaturated ones occur in vegetable oils. Two essential fatty acids must be provided by food, namely linoleic acid and linolenic acid. These are present in vegetable oils, dry fruits, nuts, and oily fish.

Metabolism of cholesterol

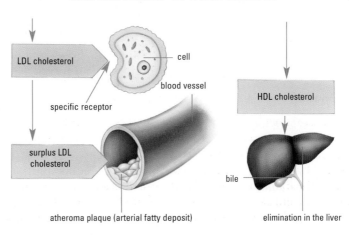

The systems of the human body

Nutrition

Vitamins

Vitamins are basic nutritional elements required in very small amounts for the satisfactory development of many of the processes of life, such as growth, immunity, various functions of the blood, the function of the endocrine system, and so on. Vitamins are divided into two major groups:

- Fat-soluble vitamins, which dissolve in fats and are absorbed with them
- Water-soluble vitamins, which are absorbed in aqueous solution.

The fat-soluble vitamins are vitamins A, D, E and K, and the water-soluble ones are vitamin C and those of the B complex.

Fat-soluble vitamins

Vitamin A

Vitamin A, or retinol, plays a part in the mechanism of vision, and is regarded as an anti-infection factor, as well as a protector of the epithelia. It occurs in nature as vitamin A in the visceral organs of animals, in fish in particular, in animal fats, egg yolk, and milk, as well as in the form of a provitamin called carotene, which is abundant in fruits and vegetables such as carrots, spinach, cabbage, apricots, or oranges.

Vitamin D

Vitamin D has a fundamental role in bone metabolism. It plays a part in the formation and calcification of bone tissue and regulates the absorption of calcium in the intestine. These functions are especially important in the periods of rapid growth, such as in infancy. Vitamin D may be of animal or vegetable origin, and can also be synthesized in the skin by the action of solar radiation. Vitamin D needs to be activated, and the liver and kidneys play a part in this. Foods rich in provitamin D are eggs, the livers of some animals, and oily fish.

Vitamin E

Vitamin E, or tocopherol, has an antioxidant effect, which helps to maintain the integrity of the cell membranes; it also plays a part in the synthesis of hemoglobin. The main sources are vegetable oils, and its absorption is associated with that of the fats.

The juice of citrus fruits is very rich in vitamin C.

Vegetable oils are the main food source of vitamin E.

Vitamins

Principal functions of the vitamins and their food sources

Vitamin K

Vitamin K plays an important part in the liver in the manufacture of substances concerned in blood clotting and its regulation. Although it is found in green vegetables (spinach, chard, etc.), it is mostly synthesized in the intestine by the action of the bacterial flora.

Water-soluble vitamins

Vitamin C

Vitamin C, or ascorbic acid, is a water-soluble acid that is concerned in the formation of collagen, the principal component of connective tissue, which is also important in the healing of wounds. It also facilitates the absorption of iron. The main sources of vitamin C are certain vegetables and fruits, especially the citrus fruits, such as oranges, lemons, and so on. It becomes inactive if boiled or if exposed to the air for prolonged periods.

Vitamin B

The B complex is made up of various different vitamins: vitamin B1 or thiamin, B2 or riboflavin, B3 or niacin or nicotinic acid, B6 or pyridoxine, and B12 or cyanocobalamin, together with folic acid, biotin, and pantothenic acid. It plays a part in the intracellular metabolism of the carbohydrates and in hematopoiesis (the formation of the blood). The complex is water-soluble and is found in whole-grain cereals, vegetables, liver, and eggs.

Vitamin A

Function
Allows for correct visual function
Protects against infections

Sources
Carrots, spinach, liver

Vitamin D

Function
Formation of bone
Absorption of calcium and phosphorus

Sources
Egg yolk, liver

Vitamin E

Function
Protects the cell membranes
Important in the fertilization of ova by spermatozoa

Sources
Vegetable fats

Vitamin K

Function
Enables blood to coagulate

Sources
Green vegetables, produced within the body by intestinal bacteria

Vitamin C

Function
Enables wounds to heal
Protects against infections

Sources
Fruits (citrus), green vegetables

Vitamin B

Function
Involved in hematopoiesis
Concerned in carbohydrate metabolism

Sources
Whole-grain cereals, liver, egg yolk

The systems of the human body

Nutrition

Minerals

Certain chemical elements play an essential part in enzymic mechanisms, bodily growth, etc. These need to be provided in food, and a deficiency or excess of them may have a variety of effects on metabolism. The most important are sodium, potassium, calcium, phosphorus, magnesium, iron, zinc, chlorine, copper, fluorine, chromium, manganese, molybdenum, and selenium.

Sodium: due to its affinity with water, sodium is responsible for maintaining extracellular water volumes. Although it can be lost in sweat or through the digestive tract, its real regulator is the kidney.

Potassium is found in the interior of cells, and is active in terms of osmosis. It is involved in the mechanism of muscular contraction and relaxation, and is regulated by the kidneys.

Calcium is found largely in the skeleton, and plays a basic role in the mineralization of bone.

Phosphorus is found in the form of phosphates, and, together with calcium, plays an important part in the formation of bone.

Magnesium is a major constituent of the skeleton, and it acts as a regulator or catalyst for a large number of intracellular reactions.

Iron is essential for the formation of the hemoglobin of erythrocytes. It is absorbed in the stomach and small intestine, and is deposited in the organs where blood cells are formed.

Zinc is involved in enzymic reactions essential for cellular reproduction, which makes it an essential mineral for growth and the regeneration of tissues.

Chlorine is very abundant in the extracellular liquids of the body, and is very closely linked to the metabolism of sodium. Its concentration is regulated by the kidneys.

Fluorine is found in bone tissue and in particular in tooth enamel, and it is therefore an essential element for dental health.

Iodine, once absorbed, is deposited in the thyroid gland, and is a constituent of the thyroid hormones.

Osmotic pressure

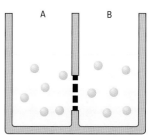

Simulation of a balanced equilibrium. There are equal concentrations of solute in A and B.

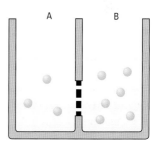

Simulation of imbalance. The A compartment has lost solute, and contains a lower solute concentration than B.

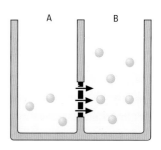

To re-establish equilibrium, water molecules move from A to B. The pressure that draws water from one compartment to the other is called osmotic pressure.

Water

Water accounts for about 60% of the body mass of an adult human being, and the body of an individual of average mass (70 kilograms) contains about 40 liters. As we have seen, it is distributed in the body between the intracellular space (the water that is present inside the cells – about 25 liters) and the extracellular space, comprising the water contained in the plasma, the lymph, the cerebrospinal fluid, and so on, as well as that in the interstitial space that surrounds the cells.

The water in these compartments moves continuously in and out of the cells across the cell membranes, depending on the concentrations of substances dissolved in it, in such a way that a balance is always maintained in the composition of the liquid in both spaces. This movement is referred to as osmosis, and the force or pressure that drives the water from one side of the membrane to the other is known as osmotic pressure.

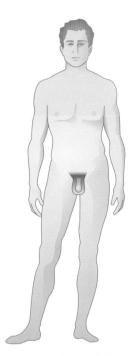

Proportion of water in the human body. Under normal conditions, water makes up 60% of the organism.

Functions of water in the body

Transport: water carries substances dissolved in it around the body.

Excretion of wastes: waste substances that cannot be reused are excreted in aqueous solution via the kidneys or in sweat.

Regulation of temperature: a third function of the water in the body is the regulation of its temperature, by the evaporation of sweat and by respiration.

Right: How water is lost from the human body (figures refer to daily losses).

Sweat: 400 ml

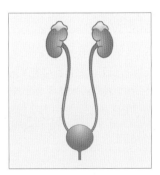
Urine: 1400 ml

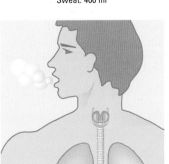
Breathing: 350 ml

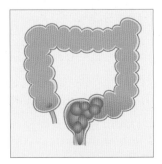
Feces: 200 ml

The systems of the human body

Nutrition

Foods

The food that individual human beings habitually eat varies considerably. In general, there are six basic groups of food recommended for daily consumption, as listed below.

1. Farinaceous products

The food wheel.

These include cereals, such as wheat, rye, barley, maize, rice, etc., and the flours derived from them. This also includes tubers, such as potatoes and yams, and vegetables like lentils, beans, peas, kidney beans, and so on. All of these are rich in carbohydrates, and are therefore good sources of energy, which is why they are often called energy foods. They are also rich in vitamins of the B complex.

Farinaceous products: tubers and legumes.

2. Fats

Fats may be of vegetable or animal origin. Among the former are olive oils, sunflower oil, maize oil, soya oil, etc., which contain unsaturated fatty acids. Animal fats are obtained from butter, bacon, egg yolk, cream, etc., which contain mainly saturated fatty acids and cholesterol.

Sources of vegetable fats.

3. Milk and dairy products

Mammalian milk is one of the most complete and healthy natural foodstuffs, and contains a high proportion of proteins, fats, carbohydrates, minerals (especially calcium), and vitamins, as well as a high percentage of water. A large number of dairy products are manufactured from milk, such as yoghurt, butter, cheese, custards, creams, and so on.

Milk, one of the most complete foodstuffs of all.

4. Vegetables

This large group of foodstuffs includes plant-derived foods cultivated in gardens, such as chard, artichokes, aubergines, marrows, onions, cabbage, spinach, and lettuce, among others. They have a high water content, a high vitamin content (especially of vitamins A and C), high mineral and fiber content, and a considerable proportion of carbohydrates in common. Generally these are known as regulator foods.

5. Meat, fish, and eggs

These provide high-quality protein, as well as fats, iron, and certain vitamins. Since the principal function of proteins is the construction of new tissues, these are known as formative foods.

The meats most frequently used for consumption are veal, beef, lamb, chicken, pork, rabbit, and game birds, as well as sausages and other meat products. Fish and other seafoods are also a source of protein, being rich in minerals such as fluorine, iodine, and phosphorus; oily fish are also important sources of polyunsaturated fats. Eggs contain a high proportion of protein in the white, and a high fat content in the yolk.

6. Fruits

Fruits such as apricots, cherries, plums, strawberries, lemons, oranges, apples, etc., are very rich in fiber, vitamins, minerals, and carbohydrates, especially sugars which are absorbed rapidly. They are also considered as regulators.

Other foods

There are other foods with characteristics of various different kinds, which have a high nutritional value. There are the dry fruits (nuts, such as hazelnuts, chestnuts, almonds, etc.), and the legumes (peas, chickpeas, and lentils).

Meat, fish, and eggs are very rich sources of animal protein.

Produce that is included in the group of cultivated vegetables, including green vegetables.

Fruits are food sources of carbohydrate, vitamins, fiber, and minerals.

The systems of the human body
Nutrition

Balanced diet

A balanced diet is one that provides the body with all its requirements, and this is achieved by eating foods each day from each of the food groups previously described. It is recommended that daily consumption should include dairy products, bread and other starchy foods, fruits and green vegetables, meat, fish or eggs, and fats, especially vegetable fats.

A balanced diet should contain: 45–65% carbohydrate, which should be divided between slow-absorption materials (starches) and rapid-absorption (greens and fruits); 15–20% proteins, which, although they should be principally of animal origin due to the better nutritional quality, may also include vegetable proteins; and 25–35% fats, in which vegetable fats should predominate, due to their higher content of unsaturated fatty acids. Food intake should be spread over at least four meals daily, in amounts which, if not exactly equal, should be largely balanced. It is advisable to plan for a breakfast that provides 25% of the total energy requirement for the day, a lunch providing for 30%, an afternoon meal covering 20%, and an evening meal providing the remaining 25%. It is important to emphasize the need for a satisfactory breakfast, which should include milk, eggs, cereals, and some fruit, and an evening meal that avoids heavy or indigestible foods.

Diet in special circumstances

Diets for pregnant and lactating women. During pregnancy, a woman will undergo a weight increase of some 12.5 kg by mid-term, which is divided between the fetus, the placenta, and the amniotic fluid, as well as the retention of bodily fluids, and will undergo a series of important changes to her metabolism. During this period she must pay attention to her specific nutritional needs, to those of the fetus in the course of growth, and to those incurred by future lactation.

Lactation diet. A breastfed infant, because of its special characteristics of physiological immaturity in terms of digestion, and because of its extraordinary capacity for growth, has dietetic requirements that are entirely different from those of an adult, and even from those of a somewhat older child.

Typical menu for a pregnant or lactating mother

BREAKFAST
Cereals with milk, bread with jam and butter, or ham and fruit.

MORNING SNACK
Fruit juice or yoghurt

LUNCH
Mixed salad, rice or pasta or potatoes or vegetables; veal or chicken accompanied by green vegetables; fruit.

AFTERNOON SNACK
Milk or yoghurt, fruit and biscuits.

EVENING MEAL
Soup or purée of vegetables or salad; fish or eggs; fruit or cheese.

A glass of milk before going to bed
Plenty of water to accompany the meals and between them

Diet

Recommended daily dietary intakes for an adult

Farinaceous foods

3–5 portions per day

Examples of one portion

60 g of bread
1 plate of rice or pasta (60 g)
1 medium-sized potato

GREENS AND VEGETABLES

2 portions per day

Examples of one portion

1 plate of salad
1 plate of fresh greens
1 tomato
1 carrot

FRUITS

2 portions per day

Examples of one portion

1 medium-sized piece of fruit
3–4 small pieces of fruit

MEAT, FISH, OR EGGS

2 portions per day

Examples of one portion

80 g of meat
80 g of fish
1 plate of legumes (beans, peas etc.)
2 eggs

MILK AND DAIRY PRODUCTS

2–3 portions per day

Examples of one portion

250 ml of milk
2 yoghurts
50 g of cheese

FATS

2 portions per day

Examples of one portion

15–20 g of oil

The systems of the human body

The cardiovascular system

The cardiovascular system carries the blood throughout the body. Central to the functioning of the system is the heart, which acts as a pump to propel the blood around the arterial system, and draws it back again through the venous system. There is also a lymphatic system which carries a fluid derived from the blood, called lymph.

The heart

The heart is a reddish-colored hollow organ that lies in the thoracic cavity between the lungs and behind the sternum, supported by the muscles of the diaphragm. It is conical in shape, with the tip pointing downwards and to the left. Inside the heart there are four chambers: two atria in the upper part, separated by the interatrial wall, and the two ventricles in the lower part, separated by the interventricular wall. The right chambers, the right atrium and the right ventricle, communicate with one another through the tricuspid valve, and the left chambers through the mitral valve.

The superior and inferior venae cavae lead into the right atrium, bringing blood from the rest of the body. The pulmonary artery leaves from the right ventricle, which delivers this blood to the lungs where it is oxygenated. The pulmonary veins, carrying oxygenated blood, arrive in the left atrium, while the aorta leaves from the left ventricle, taking blood back through the arterial system again.

The walls of the heart have three layers: the endocardium, a very thin internal membrane; the myocardium, an intermediate layer of striated muscular tissue, which is thick in the ventricular wall areas and thin in the atrial walls; and the pericardium, a membrane that envelops the heart. The movements of the myocardium are entirely automatic and involuntary.

The heart has a circulation of its own, which provides it with oxygen for its constant movement. It is known as the coronary circulation; it consists of the right and left coronary arteries, which branch off from the aorta, and blood vessels that spread throughout the heart.

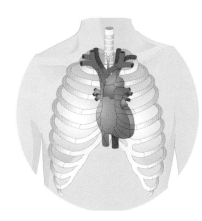

Projection of the heart on to the thoracic wall.

Anatomy

Front view of the heart

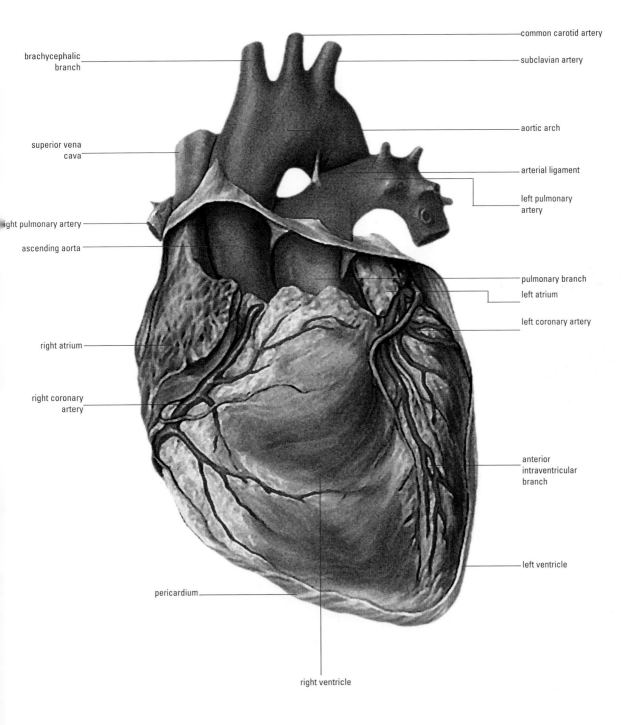

The systems of the human body
The cardiovascular system

The arterial system

The arterial system consists of a network of blood vessels called arteries, which start from the heart and extend throughout the body, carrying by means of the arterial blood the oxygen that is essential for the cells to function. The further these vessels are from the heart, the narrower they become. They are initially known as arteries, then arterioles, and finally capillaries, which are of microscopic size. It is at this final level that the transfer of oxygen to the tissues takes place.

The aorta and the arterial system

The largest and most important artery in the body is the aorta, which leaves the left ventricle upwards, and describes a curve known as the aortic arch, after which it starts to descend.

Structure of the arteries, veins, and capillaries

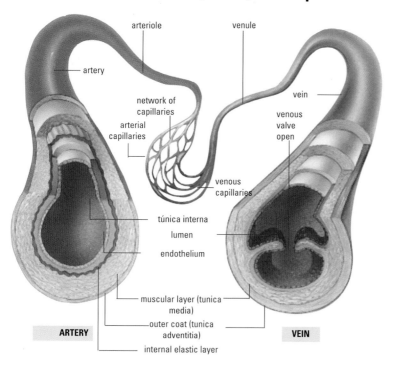

The diameters of the arterial vessels vary over their course, decreasing from the point at which they leave the heart; they are initially referred to as arteries, then as arterioles, and finally as capillaries.

Anatomy

The arterial system

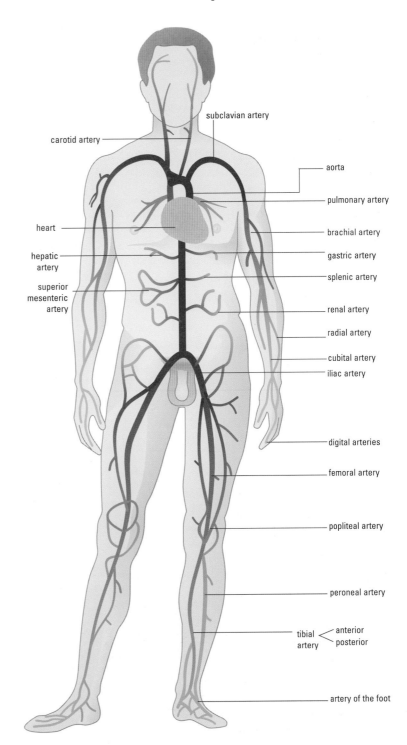

Starting from the heart by rising and then descending, the aorta first passes through the thoracic cavity and then into the abdomen (the thoracic and the abdominal aorta). The abdominal aorta forks at the pelvis into the two iliac arteries, which in turn divide into smaller arteries, each smaller than the preceding one.

From the aortic arch, arteries branch off upwards towards the head, the neck, and the upper extremities – the carotid and subclavian arteries for example. Minor branches run throughout the area, in the form of branches such as the lingual, facial, and orbital arteries.

The coeliac artery (with branches to the spleen, liver, and stomach) branches off from the abdominal aorta, and the superior and inferior mesenteric arteries (which supply the pancreas, intestines, etc.). The renal arteries, which supply the kidneys, also derive from the abdominal aorta.

Pulmonary arterial system

The pulmonary artery leaves the right ventricle and then divides into two branches, the right and left pulmonary arteries, which enter the lungs and spread out to form a similar structure to the bronchial "tree" (page 99), ending as alveolar capillaries.

The systems of the human body
The cardiovascular system

The venous system

The venous system consists of a network of blood vessels that approximately parallels the structure of the arterial network but runs in the opposite direction. These collect the deoxygenated blood, loaded with waste substances (venous blood), to the right-hand chambers of the heart, where it is passed on to the pulmonary vessels to be oxygenated and converted into arterial blood.

The walls of the veins are less elastic and muscular than those of the arteries, because the blood circulating through them is being drawn in by the suction effect of the heart. Inside the veins, any backflow of blood is prevented by a system of valves. The two major veins of the body are the superior and inferior venae cavae, two large intrathoracic

Fragment of vein, with the upper part removed

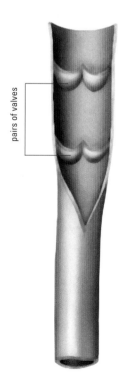
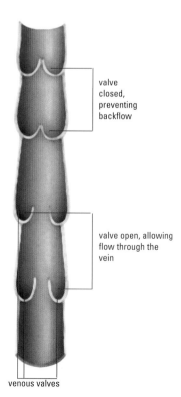

pairs of valves

valve closed, preventing backflow

valve open, allowing flow through the vein

venous valves

The blood vessels that form the venous system start as extensions of the arterial capillaries, once they have discharged the oxygenated blood. These venous capillaries merge together and form the veins, which have a substantially larger diameter.

Anatomy

Venous system

veins; the former receives the venous circulation from the upper extremities, the head, and the neck, and the latter receives the blood from the lower extremities and the abdominal and thoracic cavities.

Both the upper and lower extremities have a double venous return system, one deep and the other superficial, which ensures the correct drainage of the extremities. The two systems communicate with each other at certain points.

The veins merge to form progressively wider vessels, and change their name depending on the area that they drain.

The jugular vein, draining the head and neck, merges with the subclavian vein, which comes from the axillary vein, to form the brachycephalic vein, which curves around the heart.

Within the abdomen, the portal vein system (consisting of the superior and inferior mesenteric veins and the splenic vein) delivers blood to the liver for detoxification.

Oxygenated blood is carried from the lungs to the left side of the heart via the four pulmonary veins, which empty into the left atrium.

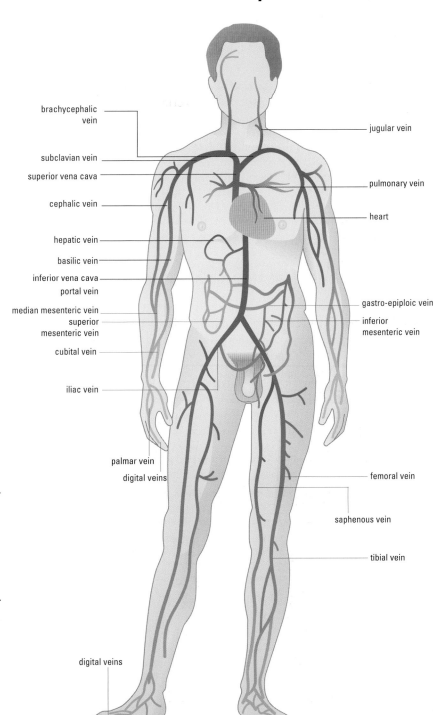

The venous system

The systems of the human body
The lymphatic system

The lymphatic system

The lymphatic system consists of a series of vessels that originate around the capillary areas where the arterial capillaries continue onwards as veins.

The lymphatic capillaries, arising at this point, carry lymph (a milky-looking liquid containing the wastes from the cellular metabolism), and discharge it via larger lymph vessels into the venous system.

Along the lymph vessels are the lymph nodes, nodules of variable volume and shape which are distributed throughout the whole of the body, and especially around the internal organs.

The lymph nodes have a largely defensive function and contain, in particular, white blood cells known as lymphocytes.

The lymph nodes act as filters, retaining micro-organisms or foreign substances that might otherwise penetrate the body. This defensive mechanism is the reason why, if there is a localized infection in any part of the body, the lymph nodes in the area increase in size.

The lymphatic vessels have valve-like folds in their inner walls, which prevent the lymph from flowing backwards. Their action is very similar to that of the veins; they finally empty into two large collectors of lymph, the thoracic duct and the major lymphatic vessel.

The thoracic duct starts in the abdominal cavity, at the junction of all the vessels coming in from the lower extremities, the intra-abdominal organs and the left-hand side of the upper part of the body. Passing through the diaphragm, it enters the thorax through the posterior mediastinum and opens into the region in which the left subclavian and left internal jugular veins meet (the jugular-subclavian angle).

The major lymphatic vessel lies in the right-hand part of the neck, and drains the lymph from the right-hand side of the upper part of the body. It empties into the region where the right internal jugular vein meets the right subclavian vein.

During their passage through the body, the lymph vessels pass through the lymph nodes, nodules of variable volume and shape which are distributed throughout the body.

Anatomy

The lymphatic system

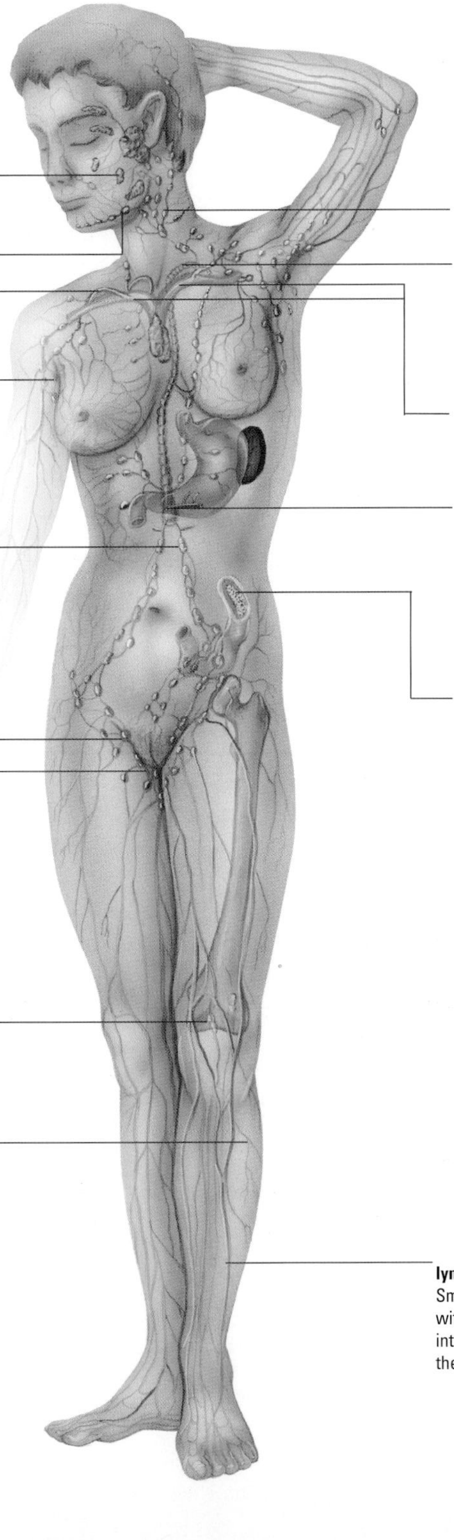

tonsil
An organ formed from a flap of lymphatic and epithelial tissue, which contains lymphatic follicles which increase in size as a response to stimulation. It produces antibodies to combat harmful organisms that might be ingested or inhaled.

submaxillary nodes

major lymphatic vessel
This receives lymph from the right arm and the right-hand side of the neck and head.

axillary nodes

lateral aortic nodes

external iliac nodes

deep lymph vessels

popliteal nodes

lymph vessels
A network of vessels extending throughout the body, into which the lymph capillaries empty their contents. They have a structure analogous to that of the veins, although their three mucosal layers are thinner and they have a larger number of valves.

cervical nodes

thoracic lymph duct
This receives lymph from the lower extremities, the thorax, abdomen, left arm, and left-hand side of the head.

subclavian veins
The lymphatic circulation discharges into these.

cisterna chyli
A vessel formed by the convergence of lymphatic vessels coming from the lower part of the body.

Peyer's patches
Clustered assemblies of lymphatic follicles in the mucosa of the small intestine.

lymphatic capillaries
Small vessels forming a network, with their ends penetrating deep into the tissues; they originate at the larger lymphatic vessels.

The systems of the human body
The cardiovascular system

Function of the heart

The cardiac cycle

In order to perform its pumping action, the heart carries out two types of movement: ventricular contraction, or systole, which dispatches the blood under pressure into the arterial system to reach all the parts of the body; and ventricular relaxation, or diastole, which allows the venous blood to flow back again into the heart.

Heartbeat: systole and diastole take place in a rhythmic and automatic manner. The sound produced during their sequence can be heard from outside the body by using a stethoscope or by placing the ear to the thorax. These sounds are referred to as the heartbeat, and correspond to the closing of the atrioventricular and sigmoidal valves.

Pulse rate: the number of heartbeats per minute is called the pulse rate. It is generally between 60 and 80, although it can vary in situations of effort or stress, or in certain disorders.

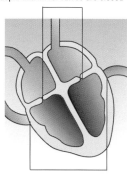

the tricuspid and mitral valves are closed

the aortic and pulmonary valves are closed

ISOMETRIC SYSTOLE
The ventricles fill with blood coming from the atria, and the tricuspid and mitral valves close.

Mechanism of contraction of cardiac muscle fibers

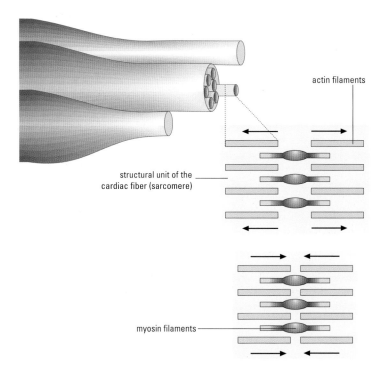

actin filaments

structural unit of the cardiac fiber (sarcomere)

myosin filaments

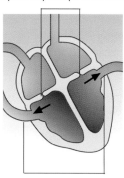

the pulmonary tricuspid valves close

the aortic and pulmonary valves open

ISOTONIC SYSTOLE
The intraventricular pressure increases and the aortic and pulmonary valves open, allowing the blood to leave the ventricles.

Regulation of cardiac movements: the command that initiates the systolic or diastolic movements of the cardiac muscle is generated in the sinoatrial node, an independent neuromuscular structure located in the right atrium, which is the physiological pacemaker of the heart. The electrical stimulus is transmitted via a network of cardiac nerves, and brings about the contraction of the myocardium. Although the sino-atrial node generates its stimuli and regulates the heartbeat in a completely autonomous and involuntary manner, the heart is still under the influence of the sympathetic nervous system, which is capable of changing the rhythm.

Mechanism of cardiac muscle contraction: the cardiac muscle is made up of myofibrils, which are composed of structural units called sarcomeres. Each sarcomere is made up of two types of filament, formed respectively of two contractile proteins, actin and myosin. These filaments are superimposed or intertwined in such a way that the interplay between them produces the contraction of the muscle.

Phases of the cardiac cycle: although systole and diastole are the basic movements of the heart, the organ functions in some ways like a four-stroke engine. The cardiac cycle starts from the moment at which the blood arrives at the atria and causes the pressure inside the heart to increase above that which pertains in the ventricles, in such a way that the right and left atrioventricular valves open and allow the blood to pass from the atrium to the ventricle, and from that point continues as follows:

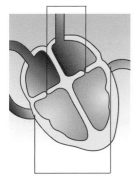

the tricuspid and mitral valves are closed

the aortic and pulmonary valves are closed

ISOMETRIC DIASTOLE
The aortic and pulmonary valves close and prevent the backflow of blood.

- **Isometric systole**: when the ventricles fill with blood coming from the atria, the pressure inside them increases. The tricuspid and mitral valves close, and prevent the blood from flowing back. The ventricles then begin to contract, and the pressure inside them automatically begins to rise.

- **Isotonic systole**: when the pressure inside the ventricles rises above that in the pulmonary artery and the aorta, the sigmoidal valves open and the ventricular blood empties into the arteries.

- **Isometric diastole**: once the ventricles have been emptied, the pressure inside them drops; the sigmoidal valves close so as to prevent the backflow of blood. The ventricles relax at this moment, and increase their volume, while still maintaining an internal pressure greater than that of the atria.

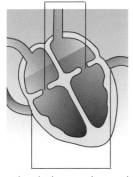

the tricuspid and mitral valves open

the aortic and pulmonary valves are closed

ISOTONIC DIASTOLE
The tricuspid and mitral valves open and again allow the blood to pass from the atria to the ventricles.

- **Isotonic diastole**: At the point of maximum relaxation, the intraventricular pressure becomes lower than the pressure of the atria, and the tricuspid and mitral atrioventricular valves then open in order to allow blood to flow from the atrium into the ventricle.

The systems of the human body

The cardiovascular system

Circulation of the blood

The blood is pumped by the heart through double circulation, namely the systemic circulation and the pulmonary circulation. The systemic circulation carries arterial blood from the left ventricle to all the tissues and returns it to the right auricle as venous blood. Conversely, the pulmonary circulation carries venous blood from the right ventricle to the lungs, where it is oxygenated, and returns it to the left auricle and thence into the left ventricle. From here it begins its passage once again through the systemic circulation.

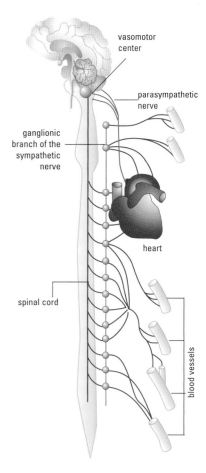

Involuntary control of the cardiovascular system.

The systemic (major) circulation

When the left ventricle contracts, the pressure within it becomes greater than that in the aorta, so that the aortic valve opens and the blood rushes through it and is forced rapidly through the whole arterial system. The movement of blood is due to the difference in pressure between the venous system and the auricle, the auricle exerting an aspiratory effect on the blood in the veins, aided by contractions of its thin muscular layer and the muscles of the extremities.

In an adult the total volume of circulating blood (volemia) is between 5 and 6 liters. Of this, some 50 to 65% is found in the venous system, which acts as a reservoir of blood for the organism. Blood distribution at a given moment depends on the degree of activity of the organ or system.

The pulmonary circulation

The contraction of the right ventricle pumps venous (deoxygenated) blood to the pulmonary artery, which carries it to the lungs. There the venous blood passes through the pulmonary arterial system until it reaches the alveolar capillaries, where gaseous exchange (O_2 for CO_2) takes place. The oxygenated blood then passes through the pulmonary venous system and is delivered by the four pulmonary veins into the left auricle. Under normal conditions the lungs may hold 10% of the total volume of blood, equivalent to approximately 500 ml.

Structure

Systemic and pulmonary circulation

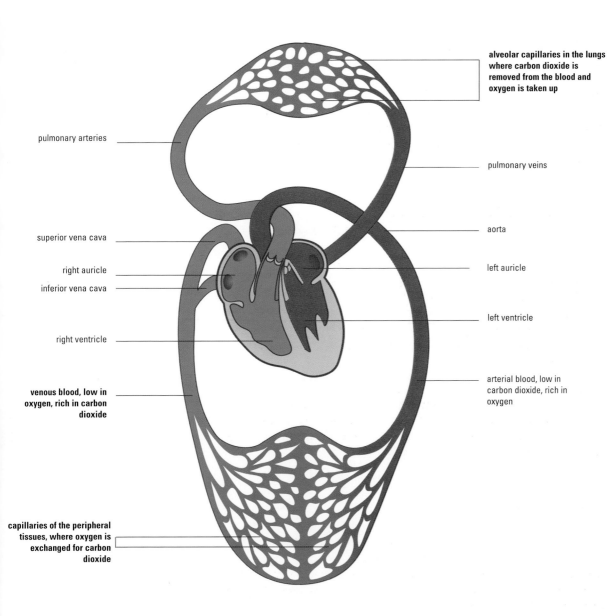

The systems of the human body
The cardiovascular system

Blood pressure

The rhythmical contractions of the left ventricle provide the force that drives the blood through the arteries. As a result there is always positive pressure in the arterial system, and this is known as arterial pressure. This arterial pressure enables the blood to reach all parts of the organism.

Arterial pressure is at its greatest at the moment of systole or ventricular contraction, and at its lowest during diastole or ventricular relaxation. These pressures are known respectively as systolic arterial pressure (maximum) and diastolic arterial pressure (minimum).

The graph of arterial pressure shows the steep upward curve during ventricular systole to the highest point—systolic pressure; it then drops again slowly to the minimum value (diastolic pressure), which coincides with ventricular relaxation.

Taking arterial pressure. Arterial pressure is conventionally measured in millimeters of mercury (mmHg); in a young adult, systolic pressure is on average about 120 and diastolic pressure about 80, although these values can vary widely according to age (generally speaking the pressure increases with age) and other circumstances.

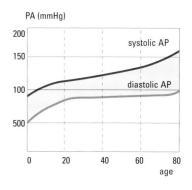

Variation of arterial pressure (AP) with age.

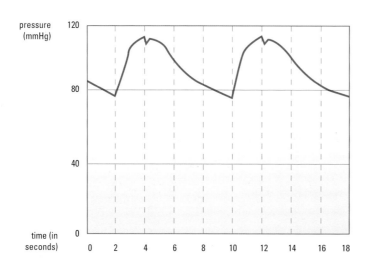

Normal pattern of the arterial pulse in the aorta

To determine arterial pressure a sphygmomanometer is used. This instrument consists of a column of mercury and a small sleeve. The sleeve is placed around the arm and inflated, thus increasing the pressure on the artery in the arm. Pressure is measured at the moment when the arterial beat is heard or ceases to be heard (this is established by feeling the pulse or using a stethoscope).

Central venous pressure. There is also pressure in the venous system, though this is lower than arterial pressure. The determining factor in venous pressure is the force with which the right auricle resists the entry of venous blood. Venous pressure can be determined by assessing the distension of the jugular veins of the neck or by using a catheter inserted into the auricle and connected to a manometer.

> In a young adult, the normal values of arterial pressure are a systolic pressure of 120 and a diastolic pressure of 80, although these figures can show wide variations according to age and other circumstances.

Central venous pressure

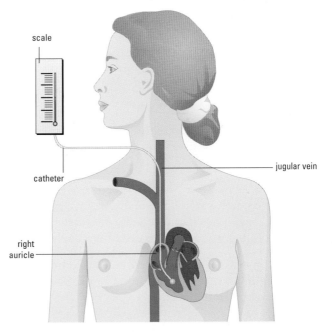

Determinación mediante la introducción de una sonda a través de una vena (en este caso la yugular) que llega hasta la aurícula derecha. Esta sonda está conectada en el exterior con una escala de valores que determina la presión en el interior de esta cavidad.

The systems of the human body

The respiratory system

The respiratory system is responsible for ensuring that oxygen from the atmosphere reaches the interior of the organism, from where it is transported by blood cells to reach every part of the body. Carbon dioxide (CO_2), a waste gas of cellular metabolism, is also eliminated by the respiratory system. In the respiratory system, air circulates in two directions: on inspiration, air rich in oxygen enters the lungs; on expiration, air containing less oxygen, but more carbon dioxide, is expelled to the outside. The vocal cords and the olfactory receptors are also found in the respiratory system.

Anatomical description

The respiratory system consists of a series of anatomical structures: the upper respiratory tract, situated in the anterior part of the face and neck, and the lower respiratory tract which lies in the neck and almost fills the thoracic cavity.

The upper respiratory tract consists of the nasal fossae and the buccal cavity which respectively lead to the upper pharynx (the rhinopharynx or cavum) and to the oropharynx. These are air passages where the incoming air is filtered, warmed and humidified. The lower respiratory tract starts at the larynx, which is a cartilaginous tube that contains the membranous folds that form the vocal cords.

The larynx continues into the trachea, which divides into the two main bronchi. These in turn lead into the lungs through what are known as the pulmonary hiluses.

The lungs are covered in a thin membrane known as the pleura. The main bronchi progressively subdivide and diminish in size to become bronchioles, very small ducts that terminate in bunches of globular structures—the alveoli—where gaseous exchange takes place.

As well as the bronchi, veins and arteries also lead into the lungs, respectively bringing venous blood rich in carbon dioxide and carrying away oxygenated blood to the arteries.

The respiratory system may be divided into the upper respiratory tract, situated in the anterior area of the face and neck, and the lower respiratory tract which lies in the neck and almost fills the thoracic cavity.

Anatomical description

General view of the respiratory system

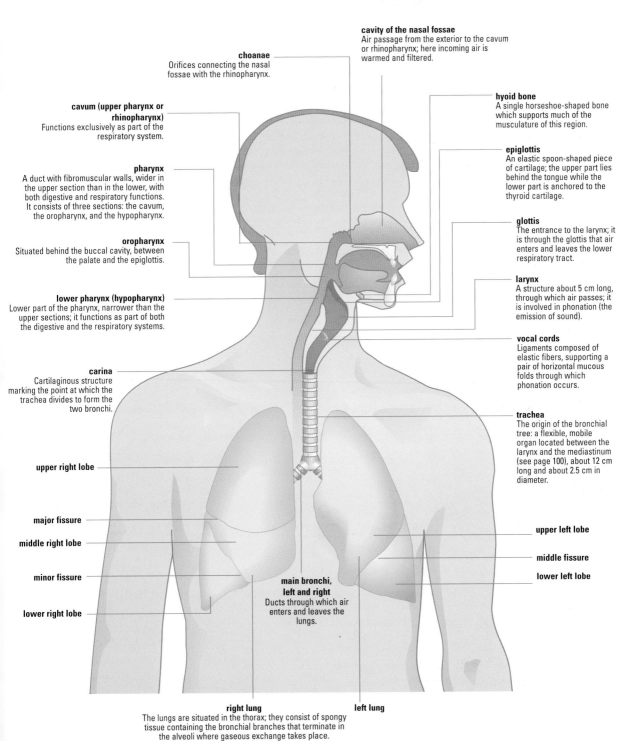

cavity of the nasal fossae
Air passage from the exterior to the cavum or rhinopharynx; here incoming air is warmed and filtered.

choanae
Orifices connecting the nasal fossae with the rhinopharynx.

cavum (upper pharynx or rhinopharynx)
Functions exclusively as part of the respiratory system.

hyoid bone
A single horseshoe-shaped bone which supports much of the musculature of this region.

epiglottis
An elastic spoon-shaped piece of cartilage; the upper part lies behind the tongue while the lower part is anchored to the thyroid cartilage.

pharynx
A duct with fibromuscular walls, wider in the upper section than in the lower, with both digestive and respiratory functions. It consists of three sections: the cavum, the oropharynx, and the hypopharynx.

glottis
The entrance to the larynx; it is through the glottis that air enters and leaves the lower respiratory tract.

oropharynx
Situated behind the buccal cavity, between the palate and the epiglottis.

larynx
A structure about 5 cm long, through which air passes; it is involved in phonation (the emission of sound).

lower pharynx (hypopharynx)
Lower part of the pharynx, narrower than the upper sections; it functions as part of both the digestive and the respiratory systems.

vocal cords
Ligaments composed of elastic fibers, supporting a pair of horizontal mucous folds through which phonation occurs.

carina
Cartilaginous structure marking the point at which the trachea divides to form the two bronchi.

trachea
The origin of the bronchial tree: a flexible, mobile organ located between the larynx and the mediastinum (see page 100), about 12 cm long and about 2.5 cm in diameter.

upper right lobe

major fissure

middle right lobe

upper left lobe

minor fissure

middle fissure

lower left lobe

lower right lobe

main bronchi, left and right
Ducts through which air enters and leaves the lungs.

right lung **left lung**
The lungs are situated in the thorax; they consist of spongy tissue containing the bronchial branches that terminate in the alveoli where gaseous exchange takes place.

The systems of the human body

The respiratory system

The nasal fossae

These cavities lie in the central part of the face. They allow air to pass to the pharynx through the choanae.

The nose is a structure formed by the nasal bones together with a cartilaginous extension, the whole region being covered by skin. It has two orifices (nostrils) in the lower section.

The nasal septum is a bony cartilaginous partition between the left and right nasal fossae.

The nasal conchae are three bony protuberances situated one above the other on the lateral walls of the nasal fossae. Their function is to create turbulence in the inspired air, thus preventing air from passing directly into the pharynx.

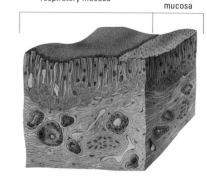

Structure of the nasal mucosa as seen through the microscope.

The paranasal sinuses are cavities in the bones surrounding the nasal fossae and connecting with them. They are eight in number and are distributed symmetrically: two frontal, two ethmoid, two sphenoid, and two maxillary. Their function is to secrete mucus and to warm and humidify the air.

The nasal mucosa (also known as the respiratory mucosa) covers the nasal fossae and contains a large number of mucosal glands. In the uppermost section there is a specific area of mucosa that is connected to the cerebral cortex. This is known as the olfactory mucosa; it picks up olfactory stimuli and transmits them to the brain for processing.

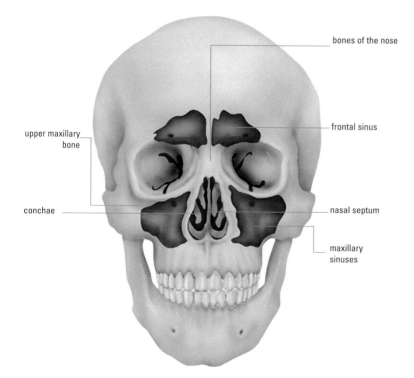

Nasal fossae: anterior view.

Pharynx

The pharynx is a duct of muscular and fibrous tissue that is involved in both breathing and digestion. It consists of three sections: the upper pharynx, also known as the rhinopharynx or cavum, the middle pharynx or oropharynx and the lower pharynx or hypopharynx.

The upper pharynx is concerned exclusively with breathing. Air from the nasal fossae passes through the choanae. In the lateral walls are the ends of the Eustachian tubes, which connect the upper pharynx with the ear cavities.

The middle pharynx is the downwards continuation of the cavum and extends to the base of the tongue, where the tonsils are located. The oropharynx has a double function: it allows either air or food to pass through it.

The lower pharynx also has a double function and is the downward extension of the oropharynx, reaching as far as the esophagus. In its anterior part it communicates with the larynx.

> The pharynx is a duct of muscular and fibrous tissue that is involved in both breathing and digestion. It consists of three sections: the rhinopharynx, the oropharynx, and the hypopharynx.

Internal structure of the nasal fossae and pharynx

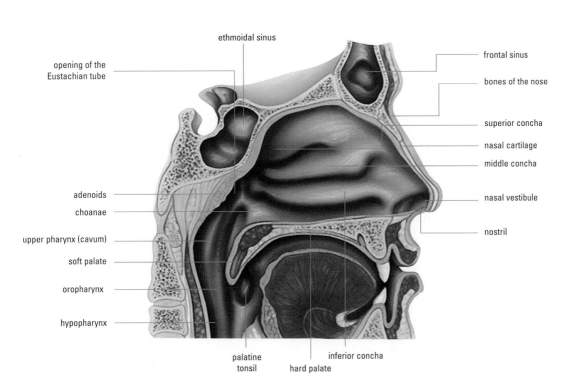

The systems of the human body
The respiratory system

The larynx

This is a duct formed of cartilaginous rings connected by ligaments and muscles. It lies towards the front of the neck and opens into the anterior lower section of the hypopharynx; its lower limit is the trachea. It is covered in a thin mucosa, which is continuous with the pharyngeal mucosa.

The epiglottis is a cartilaginous structure that closes the opening between the hypopharynx and the larynx, preventing food from entering during swallowing and allowing the passage of air during inspiration.

The glottis is the region in which the vocal cords are located. These divide the glottis into two parts, known as the superior glottis and the inferior glottis. The latter narrows to become the subglottal space, which is continuous with the trachea.

The vocal cords are folds of membranous tissue situated in the middle section of the glottis. The superior and inferior vocal cords are separated by a small cavity known as Morgagni's ventricle. The vibration of these cords as a result of the passage of air is what produces the voice.

The laryngeal cartilages form the basis of the walls of the larynx and are, from top to bottom:

- The epiglottal cartilage, which forms the basis of the epiglottis. It is oval in form, ending in a sharp downward-pointing tip. Its rocking movement enables it to open or close the larynx.

- The thyroid cartilage is formed by two interconnected rectangular laminae with an open angle to the posterior, similar to an open book with a spine protruding at the anterior of the neck; this is the protuberance known as the Adam's apple.

- The cricoid cartilage is situated below the thyroid cartilage; it is the only cartilage that forms a complete ring.

- The arytenoid cartilages are two triangular plates lying behind, and linking, the cricoid and thyroid cartilages.

- The corniculate and cuneiform cartilages are small cartilages, which together complete the articulation of this complex mechanism.

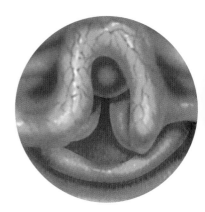

The open epiglottis and larynx (below), as seen through an endoscope.

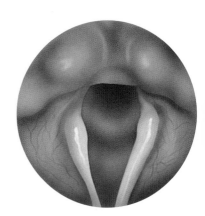

An endoscopic view of the vocal cords.

Structure

Front view

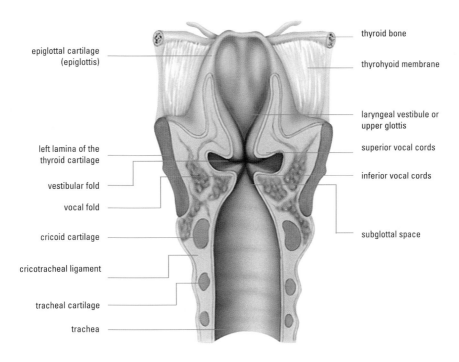

Side view

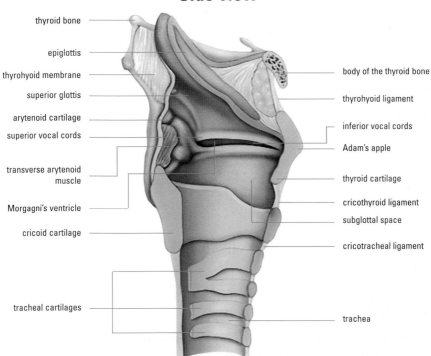

The systems of the human body

The respiratory system

The trachea

The trachea is a tubular structure, 11 or 12 cm long in an adult, which extends downwards from the larynx deep into the thorax. Here it occupies the posterior section of the mediastinal cavity, together with the esophagus and the thoracic aorta. It ends in a bifurcation known as the carina, from which the two main bronchi, left and right, arise.

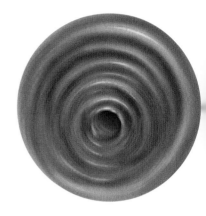

An endoscopic view along the trachea.

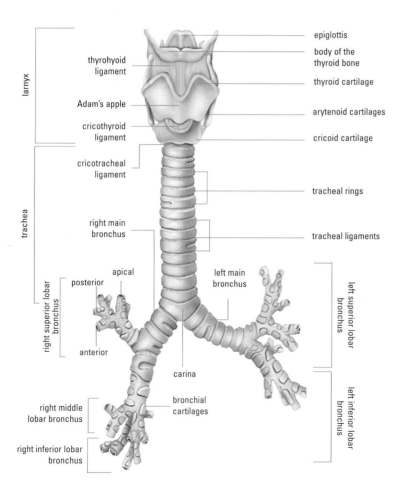

External view of the larynx, trachea, and bronchi.

Structurally it consists of a series of some 16 to 20 rings of cartilage, connected by ligaments. The "rings" are incomplete, with a gap on their posterior side where smooth muscle fibers complete the circle. The entire structure is thus flexible enough to contract and dilate. The trachea is lined by a mucosa carrying numerous ciliated cells and mucus-secreting glands.

Anatomy

Larynx, trachea and bronchi (posterior view)

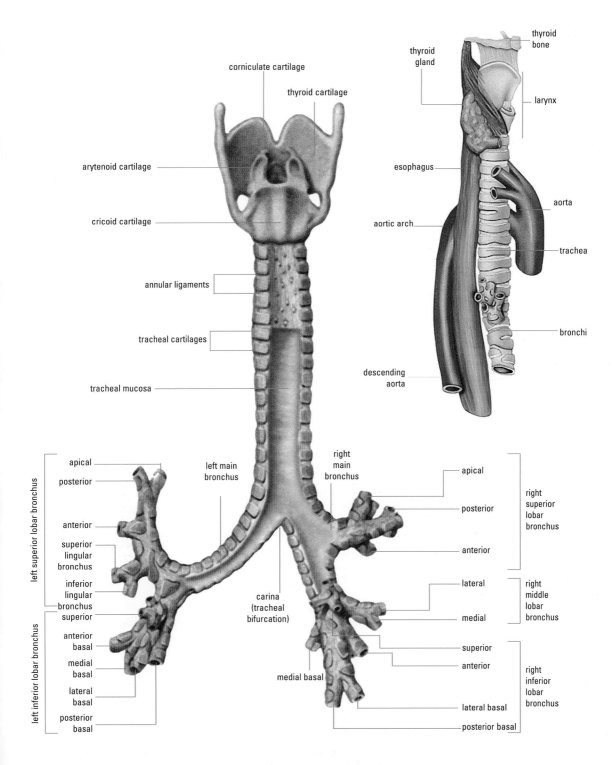

The systems of the human body
The respiratory system

The bronchi

The lowest section of the trachea divides into two branches, each of which goes to one of the lungs. These are the left and right main bronchi, and their structure is similar to that of the trachea.

Just below the carina, each bronchus enters the appropriate lung through an opening known as the pulmonary hilus, through which the pulmonary arteries and veins also enter.

Once inside the lungs, the main bronchi continue to subdivide, each division becoming progressively smaller, to form the lobar bronchi, each of which corresponds to one lobe of the lung. These subdivide into the segmental bronchi, which extend into each segment of the lung, and finally the lobulilar bronchi and the bronchioles, which are the narrowest, smallest extensions of the bronchial tree.

The bronchioles terminate in bunches of microscopic sacs known as the pulmonary alveoli. It is in these minute cavities that the exchange of gases that constitutes the basis of respiration takes place. It is estimated that the lungs contain hundreds of millions of these tiny alveoli.

The patterns of bronchial division are not the same in both lungs, nor are the divisions of the lungs themselves identical. The right main bronchus follows a more vertical path than it does on the left and after passing through the pulmonary hilus it divides into three lobar bronchi: superior, middle and inferior. These then divide into the corresponding segmental and lobulilar bronchi and bronchioles.

The left main bronchus enters the left lung more obliquely and produces only two lobar bronchi: the superior and inferior. Throughout its length the structure of the bronchial tree is very similar to that of the trachea, being composed of a series of cartilaginous rings joined by ligaments and surrounded by muscles that allow dilation and constriction.

An endoscopic view of the bronchi.

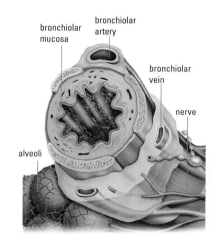

Internal structure of a bronchiole.

Anatomy

Anatomical diagram of the bronchial tree

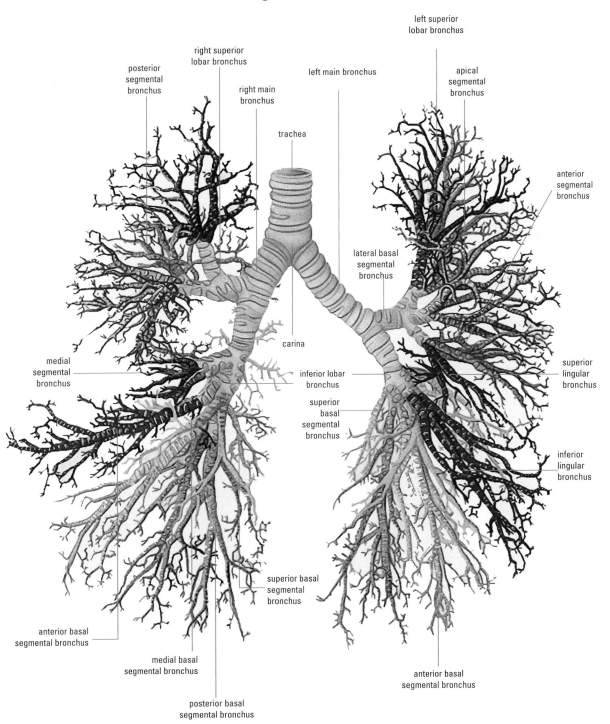

The systems of the human body
The respiratory system

The lungs

The lungs are two organs of spongy tissue covered in a thin membrane (the pleura) and enclosed within the ribs. They are situated on either side of the thoracic cavity, almost completely filling it.

The cavity between the lungs is known as the mediastinum; it contains, among other structures, the heart. On their external, convex side the lungs make contact with the ribs and on their smooth inner side with the structures of the mediastinum. Their angled base rests on the diaphragm, the muscle forming the lower limit of the thoracic cavity. At the center of the internal sides of the lungs are the pulmonary hiluses, the point of entry of the bronchi, arteries, veins and lymphatic vessels.

The drawing below shows how the lungs are divided into lobes by fissures. There are two in the right lung, creating three lobes, and one in the left lung, creating two lobes—the superior and the inferior. Each of the pulmonary lobes is in turn divided internally into various segments.

The lungs consist of pink connective tissue, which has a soft consistency and great elasticity, and which surrounds and encloses the bronchi, bronchioles, and alveoli.

Segmental and lobar division of the lungs

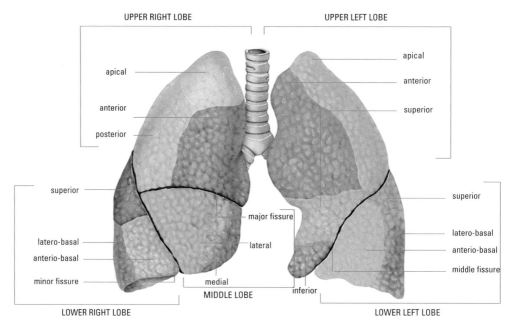

Anatomy

View of the lungs

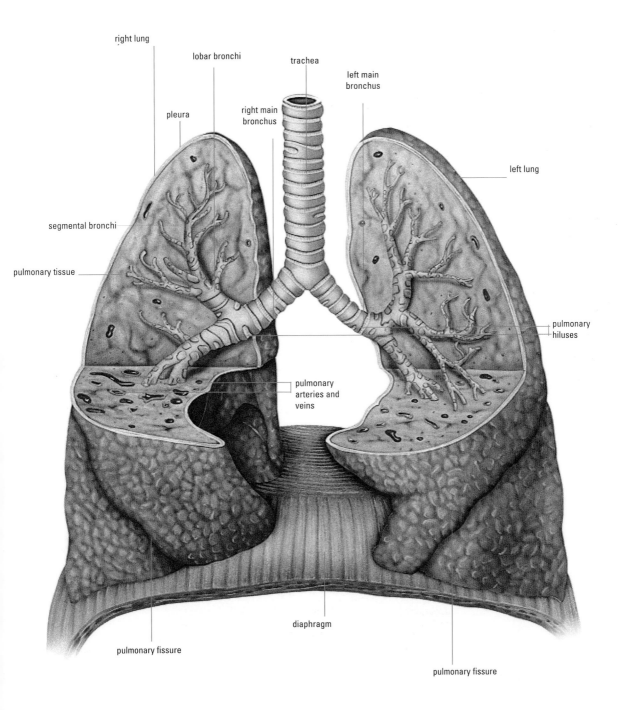

The systems of the human body
The respiratory system

The pulmonary blood vessels

After leaving the heart the left and right pulmonary arteries enter the lungs via the pulmonary hiluses and from there branch into a network in a similar manner to that of the bronchi, terminating in fine arterial capillaries in the walls of the alveoli. Here the arterial capillaries become venous capillaries, which gradually join up together to form the pulmonary veins, two to each lung. These leave the lungs through the hiluses and lead back to the heart.

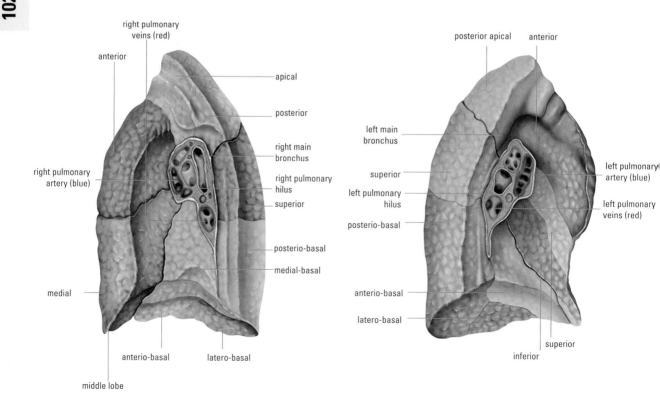

Thoracic muscles

> These muscles contract and relax involuntarily, thus ensuring that breathing continues even during sleep.

The functioning of the whole complex respiratory system depends on the action of the thoracic muscles. This action takes place in two modes: inspiratory, which enables air to enter, and expiratory, which helps to expel it. The principal muscles of inspiration are the diaphragm and the external intercostal, sternocleidomastoid, scalene, and major serratus muscles. The effect of the contraction of any of these is to increase the size of the thoracic cavity.

The principal muscles of expiration are the abdominal, the internal intercostal, and the posterior serratus muscles. These muscles contract and relax involuntarily, thus ensuring that breathing continues even during sleep.

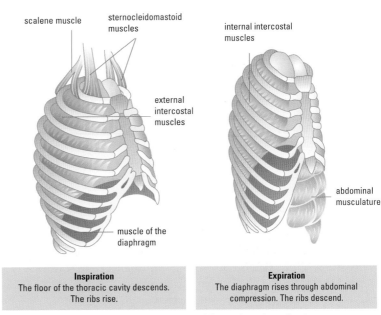

Inspiration
The floor of the thoracic cavity descends. The ribs rise.

Expiration
The diaphragm rises through abdominal compression. The ribs descend.

Thoracic movement during breathing and the musculature that produces it.

The pleura

The entire outer surface of the lungs is covered by a double membrane known as the pleura. The internal or visceral pleura is in contact with the lung and the external or parietal pleura is attached to the rib cage and diaphragm. There is also a mediastinal pleura; this covers the organs of the mediastinum, isolating them from the lungs.

The systems of the human body
The respiratory system

Functions

The principal functions performed by the respiratory system are to bring oxygen to the blood and to remove carbon dioxide from it by the process of breathing, or ventilation. Other functions of the respiratory system that are also important to human life are phonation and olfaction.

Breathing

Breathing is a complex process, the purpose of which is to convey oxygen from the air around us to the pulmonary alveoli; here this oxygen passes into the blood and the carbon dioxide produced during cellular metabolism passes into the alveoli and is eventually expelled into the air. This process is repeated automatically throughout the life of a human being and consists of different phases.

Inspiration is the first phase of breathing. Air from outside passes into the lungs, eventually reaching the alveoli. For this to happen the pressure within the lungs must be lower than the pressure outside, thus enabling air to be drawn inwards. The difference in pressure is achieved by means of thoracic expansion, which takes place mainly through the action of the intercostal muscles and the diaphragm.

During the process of inspiration, the respiratory system warms, humidifies, and filters the incoming air, to ensure that it arrives in optimum condition. Involved in this process are the nasal fossae, the paranasal sinuses, the larynx, the pharynx, and the trachea.

Alveolar diffusion: the exchange of oxygen and carbon dioxide takes place in the pulmonary alveoli, which are surrounded by thin-walled blood capillaries. This exchange takes place through the passive diffusion of the gases through the alveolar and capillary walls; the oxygen, which is in high concentration in the alveolar air, passes into the venous blood in the capillaries and the carbon dioxide passes from the capillary blood into the alveoli. The capillary blood, now charged with oxygen, leaves the lungs via the pulmonary veins and, after passing through the heart, is distributed throughout the organism in the form of arterial (oxygenated) blood.

> During the process of inspiration, the respiratory system warms, humidifies, and filters the incoming air, to ensure that it arrives in optimum condition. Involved in this process are the nasal fossae, the paranasal sinuses, the larynx, the pharynx, and the trachea.

Process

Gaseous exchange in the alveoli

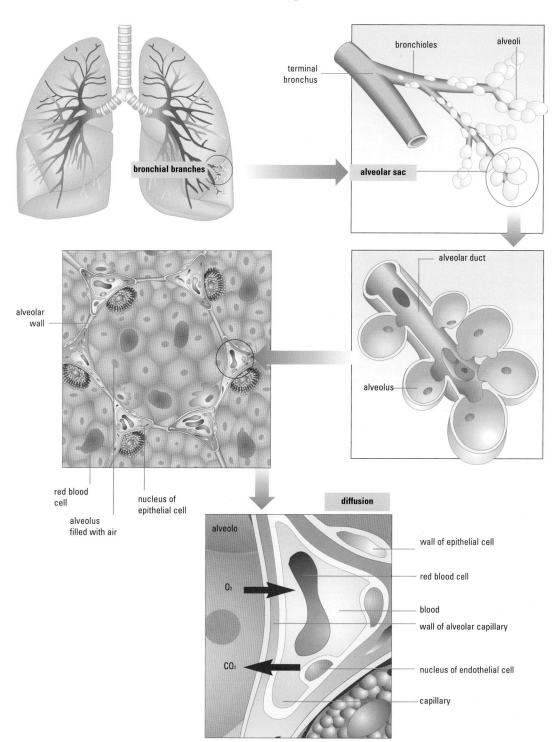

The systems of the human body

The respiratory system

Expiration (exhalation) is the process of the expulsion of alveolar air after gaseous exchange has occurred, and results from the relaxation of the inspiratory muscles. When these muscles stop contracting, the elasticity of the pulmonary tissue enables the lungs to return to their initial position, thus expelling the air contained within. When a deliberate act of expiration is required, the contraction of the abdominal muscles pushes the diaphragm upwards, bringing about the supplementary expulsion of air.

Control of breathing: in normal circumstances the processes of breathing are repeated 15 to 20 times a minute (normal rate of respiration). In newborn and nursing infants the rate of respiration is higher than in adults.

The muscular activity needed for breathing is controlled by the neurological center of respiration in the medulla oblongata of the brain. This center of respiration functions automatically and involuntarily, thus ensuring that the process continues during sleep, loss of consciousness, etc. In addition to this involuntary control, there is also a degree of conscious control of breathing which allows it to be speeded up or slowed down; the process can even be halted for a limited period.

Control of breathing

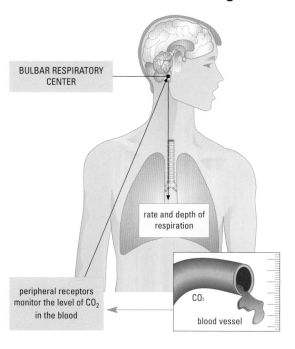

The center of respiration receives information about the concentration of carbon dioxide in the blood from peripheral receptors and adjusts the rate and depth of respiration accordingly. It may therefore be said that blood carbon dioxide levels indirectly control respiration.

Phonation

Phonation is the result of the passage of air from the lungs through the larynx where the vocal cords are situated. It is the vibration of the vocal cords that produces the voice. The characteristics of the voice depend on several variables: the volume of the voice depends on the amount of air passing through the vocal cords; the tone depends on the degree of tension in the vocal cords; and the timbre on the effect of resonance in the paranasal sinuses, the larynx, and the thoracic cavity. As a result the voice of any individual has characteristics that differentiate it from every other.

The neurological center for speech is situated in the brain. This center controls both the emission of sounds by the vocal cords and also the articulation of these sounds to convert them into syllables and words. Articulation takes place in the buccal cavity through the action of the tongue, the lips, and the palate, which produce nasal, occlusive, fricative, affricative, and liquid sounds.

DIFFERENT BUCCAL SOUNDS

Nasal	m, n
Occlusive	p, t, k, b, d, g
Fricative	f, s, z, x
Liquid or lingual	l, r

Olfaction

Olfaction, or the ability to perceive the odors of people, animals and objects, also takes place within the nasal fossae. This faculty is of vital importance in some species but is used rather less by humans.

This function will be examined in greater detail in the section on the organs of the senses (page 162–171). At this point it is sufficient to note that it takes place inside the nasal fossae in an area of mucosa (the olfactory mucosa) that has a close nervous connection with certain higher neurological centers.

Mechanisms of voice production

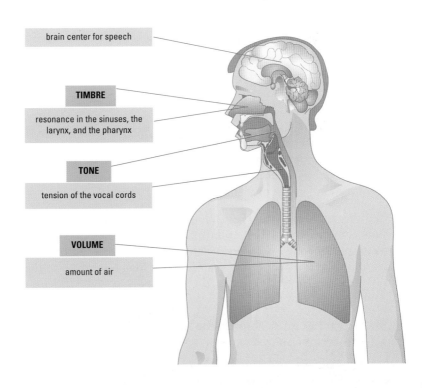

- brain center for speech
- **TIMBRE** — resonance in the sinuses, the larynx, and the pharynx
- **TONE** — tension of the vocal cords
- **VOLUME** — amount of air

The systems of the human body

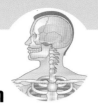

The musculoskeletal system

The bony skeleton

The human skeleton, like that of other vertebrates, is internal to the muscles, which are inserted into it. It has bilateral symmetry, which is to say that it is symmetrically distributed about an axis, in this case a longitudinal axis running through the spinal column.

The basic element of the skeleton is bone, composed of bone tissue. Bone tissue is a special type of connective tissue that combines the rigidity required by any supporting element with the plasticity needed to allow growth and contribute to the flexibility of the skeleton. Under the microscope, bone tissue is seen to be highly vascularized and formed of cells and calcified extracellular components; it is these that give the skeleton its characteristic strength.

The cells of bone tissue are known as osteoblasts and osteoclasts. Osteoblasts are responsible for forming the organic matrix of the tissue; this later mineralizes, leaving the osteoblasts enclosed within it. The osteoclasts are the cells responsible for the resorption and remodelling of bone. In a healthy individual a perfect balance is achieved throughout life between the formation (osteoblasts) and destruction/remodelling (osteoclasts) of bone tissue.

Types of bone

The skeleton consists of about 200 individual bones distributed in the central area or axial skeleton (the cranium, spinal column, sternum and ribs) and in the peripheral area or appendicular skeleton (the bones of the limbs, scapulae, clavicles and pelvis). Detailed views of the bones of the human body are shown on pages 109 and 111.

With regard to internal structure, bone can be of two types: compact bone, formed by parallel columns of concentric layers of bony tissue; and spongy bone formed by irregular interwoven beams of bony plates known as trabeculae. For example, in a long bone such as the femur, the spongy bone is located at the ends (epiphyses) and the compact bone in the central part (diaphysis).

With regard to external form, the bones of the skeleton may be long, like the femur, short like the vertebrae, or flat like the scapula.

Different types of bone

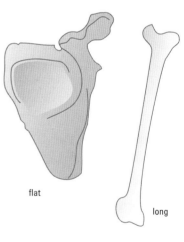

flat

long

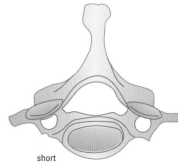

short

Structure

Anterior view of the skeleton

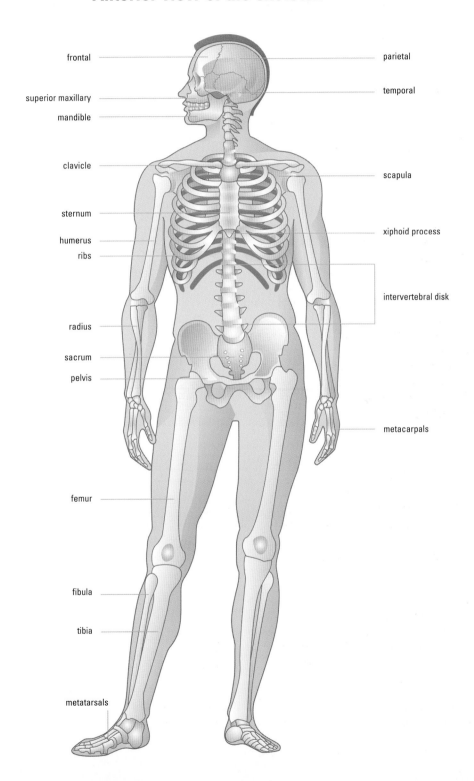

The systems of the human body
The musculoskeletal system

The joints

The bones are interconnected by articulations or joints, which allow different degrees of movement according to their structure, and which together give the skeleton mobility and flexibility. Some, like those of the bones of the cranium (known as sutures) allow scarcely any movement; others, such as the vertebrae, allow a small degree of movement while others, such as the hip joint, permit much more.

Generally speaking, a typical movable joint consists of the following elements:

The articular cartilage is a special type of cartilage that covers the surface of the bone involved in articulation.

The synovial membrane lines the joint as far as the border with the articular cartilage. It contains synovial fluid, which performs the extremely important role of lubricating the joint.

The joint capsule is located outside the synovial membrane. It consists of a small sleeve of fibrous connective tissue, which holds the bones of the joint together.

The ligaments are strips of very strong connective tissue which are inserted into the bones of the joint. They are located outside the joint capsule and have the function of allowing certain movements of the joint while restricting others.

In addition some joints have special elements such as meniscuses which are small cushions of connective tissue situated between the articular cartilages and which allow the surfaces to move smoothly against each other. There are meniscuses in the joints of both the knee and the jaw.

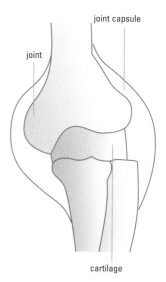

Cartilage

Cartilage consists of relatively hard, whitish material, rich in elastin and collagen fibers and with considerable flexibility. Skeletal structures which, because of their function, need to be more movable than bone are made of cartilage. Cartilages can be classified according to the proportion of collagen and elastin fibers they contain: hyaline cartilage (e.g. the nasal septum, articular cartilage), elastic cartilage (e.g. the ear), and fibrous cartilage (e.g. the intervertebral disks).

Structure

Posterior view of the skeleton

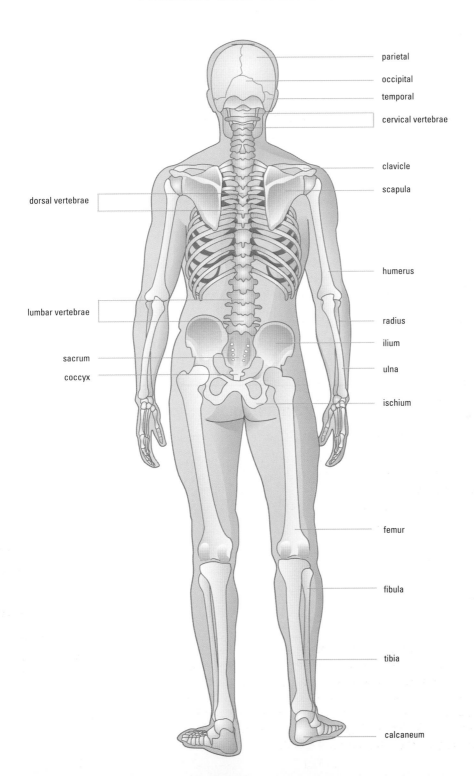

The systems of the human body
The musculoskeletal system

The muscles

The muscles lie beneath the skin and are in close relationship with the skeleton. There are more than 600 muscles in the human body, and their role is to provide movement to the rest of the body via their insertions at key points in the bones. At these points muscle tendons, very strong strips of connective tissue, connect the muscle and the bone. The psoas muscle, for example, which runs from the lumbar vertebrae to the inner side of the femur, lifts the thigh. Detailed views of the muscles of the human body are shown on pages 113 and 115.

Muscles that are inserted into the bones are known as skeletal muscles and are the most numerous. There is also muscle tissue – visceral muscle – in the walls of the blood vessels and certain viscera such as the gut, the uterus and the urinary bladder. Finally, there is the muscle tissue of the heart, known as cardiac muscle or myocardium.

The muscles, of which there are more than 600, are attached to the bones by means of tendons inserted into the bones, and are usually arranged in opposing groups.

Cross-section of an artery showing the layer of smooth muscle

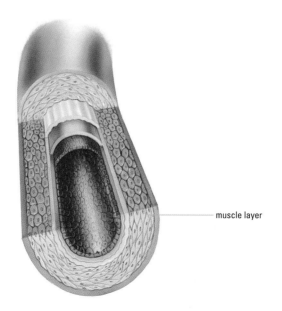

muscle layer

Muscles inserted into the bones are known as skeletal muscles, though there are other types of muscle, such as visceral muscle in the walls of the blood vessels, the uterus, and other organs, and also cardiac muscle or myocardium.

Structure

The muscles: anterior view

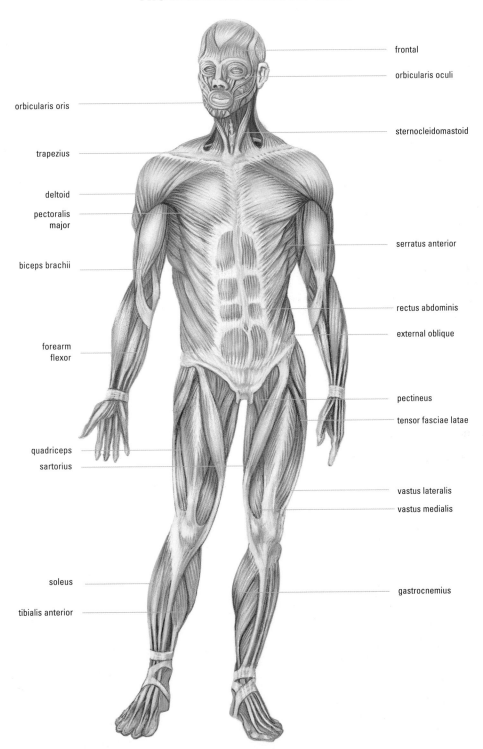

The systems of the human body

The musculoskeletal system

Types of muscle

Muscle types may be classified according to their shape: ring, fusiform (with one or more bulges), quadrilateral, triangular, etc.

They may also be classified according to the direction of the muscle fibers: parallel, oblique and spiral.

If we look at the internal structure we find two types: striated (striped) and smooth. Striated muscle is found in skeletal muscle, as described previously, and also in cardiac muscle. It is so named because the muscle fibrils are arranged in such a way that they appear like stria or grooves along the length of the muscle surface. Smooth muscle is found in visceral muscle. The smooth arrangement of the fibers (myofibrils) gives it its name.

Skeletal striated muscle is *voluntary*, that is to say it is under our conscious control. On the other hand, striated cardiac muscle and smooth muscle are *involuntary*, that is to say that they function independently of conscious control.

The contraction of the cardiac muscle or myocardium deserves special mention. This takes place as a result of involuntary electrical impulses generated by specialized cells located in the heart itself, in the sinoatrial and auriculo-ventricular nodes.

Muscular contraction

Skeletal muscles consist of groups of long cells known as myocytes or muscle fibers, which are packed very closely together. Bundles of these cells together form muscle fascicles, and the muscle itself consists of a group of these fascicles or bundles. Muscle fibers consist of many smaller fibers, known as myofibrils, each of which is composed of a number of contractile myofilaments—some fat, some thin—consisting basically of two specialized proteins: actin and myosin.

Muscular contraction takes place by the sliding of the fat myofilaments along the thin myofilaments, thus reducing the overall length of the muscle. This sliding action takes place as a result of interaction between the actin and myosin molecules. The stimulus that generates skeletal muscular contraction is electrical and is produced voluntarily in the brain.

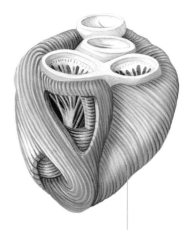

myocardium

Structure of the walls of the heart, showing the substantial muscular content.

Structure

The muscles: posterior view

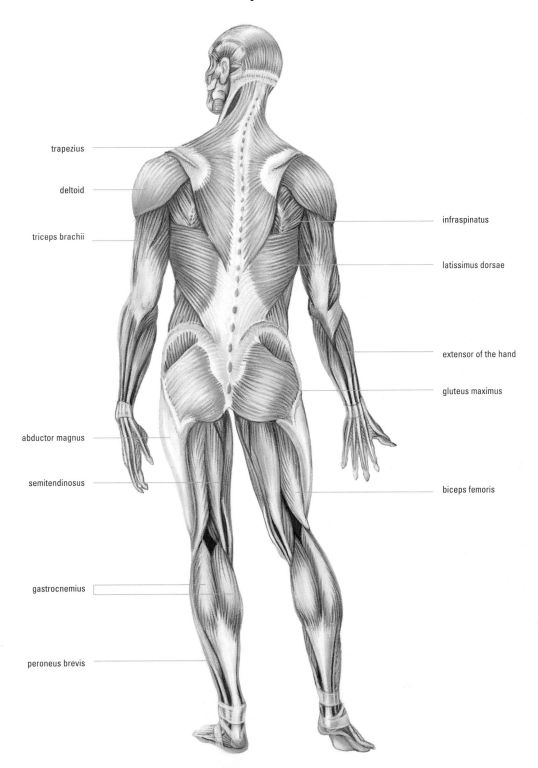

The systems of the human body

The reproductive system

The reproductive system consists of the organs that enable human beings to create new life and to reproduce themselves. The human species is divided into two sexes: male and female. Each individual produces basic sex cells; when male and female sex cells unite, the product will develop into a new individual.

Spermatozoon

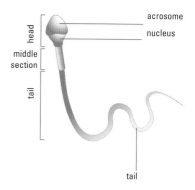

Male genital organs

The most important of the male genital organs are the penis and the testicles. The penis is cylindrical in form and consists of three parts: the base, the body, and the glans. Inside the body of the penis are the corpora cavernosa and the corpus spongiosum, which fill with blood during erection. The urethra, from which urine and semen are expelled, passes through the penis. The glans is the most sensitive area of the penis. At its center is the urethral orifice, the point of exit of the urethra. The principal function of the penis is to deposit sperm inside the vagina of the woman during the act of coitus.

The testicles are a pair of glands that produce the spermatozoa and hormones that control the sexual life of the male.

Anatomy

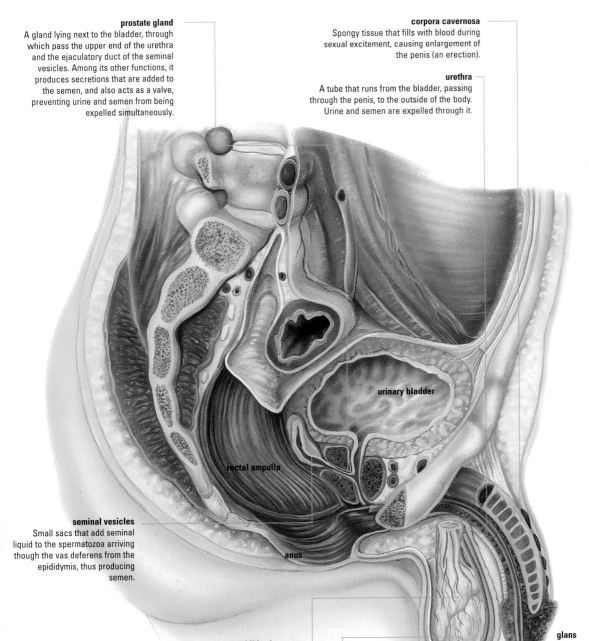

prostate gland
A gland lying next to the bladder, through which pass the upper end of the urethra and the ejaculatory duct of the seminal vesicles. Among its other functions, it produces secretions that are added to the semen, and also acts as a valve, preventing urine and semen from being expelled simultaneously.

corpora cavernosa
Spongy tissue that fills with blood during sexual excitement, causing enlargement of the penis (an erection).

urethra
A tube that runs from the bladder, passing through the penis, to the outside of the body. Urine and semen are expelled through it.

urinary bladder

rectal ampulla

seminal vesicles
Small sacs that add seminal liquid to the spermatozoa arriving though the vas deferens from the epididymis, thus producing semen.

anus

epididymis
A spirally-twisted bundle of tubes, situated behind the testicles and linked by the vas deferens to the seminal vesicles. Here spermatozoa mature and are stored prior to ejaculation.

testicle
An oval-shaped gland where the male hormone (testosterone) and spermatozoa are produced.

scrotum
Sac containing the testicles and the epididymis.

glans

urethral meatus

The systems of the human body
The reproductive system

Functions of the testicle

The testicle produces spermatozoa and male hormones that distinguish and control the sexual life of the male.

The testicle is composed of two elements: groups of Leydig cells, which produce the male hormones, and a system of seminiferous tubules where sperm is produced and through which it is transported. The testicle, like the ovary in the female, therefore has a double function: firstly, the secretion of hormones and, secondly, spermatogenesis (the formation of spermatozoa).

Endocrine function of the testicle

The endocrine function of the testicle consists of the production of male sex hormones. Leydig cells play a key role in hormone synthesis, which takes place under the influence of the hypothalamus and the pituitary gland.

Leydig cells only represent between 1 and 10% of total testicular volume. They are polygonal in form and are distributed between the seminiferous ducts or tubes, forming groups. Their function is, as we have seen, to secrete male sex hormones or androgens (especially testosterone) and also some estrogen. The hormones pass into the blood and function in several ways. They control the differentiation of male sex characteristics during puberty, such as the growth of body hair and the development of muscle mass. They also control spermatogenesis.

Exocrine function of the testicle

The testicle is divided by partitions known as septa into some 250 lobules of fibrous tissue, which arise from the testicular or Haller network. Each lobule contains between one and four twisting tubules, each about 60 cm long. It is in these tubules that spermatogenesis takes place.

Spermatogenesis occurs in two phases: spermatogenesis proper – the phase of cellular division – and spermiogenesis, the phase of cellular maturation.

In spermatogenesis proper, the roles played by Sertoli cells, Leydig cells, and the hormones testosterone and follicle-stimulating hormone (FSH) are of prime importance.

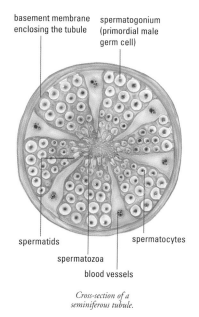

Cross-section of a seminiferous tubule.

The testicles lie on the left and right of the penis and are enclosed in a sac called the scrotum. Spermatozoa formed in the testicles pass into the epididymis which is attached to the postero-lateral area of the testicle.

Function

Hormonal regulation of testicular function by the cerebro-testicular axis: negative feedback mechanism

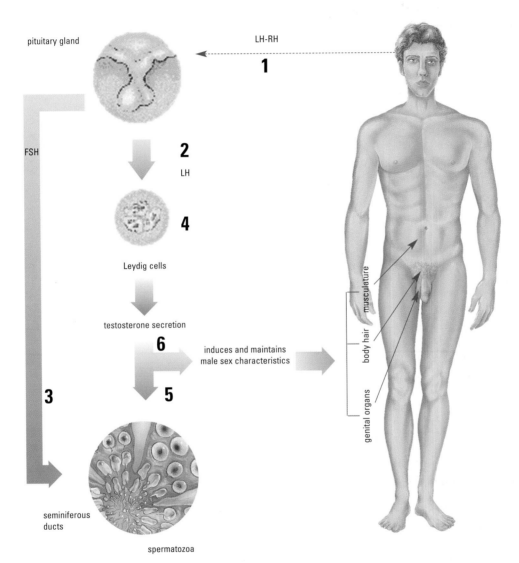

1. The hypothalamus secretes the hormone LHRH or Gn-RH (the hormone that stimulates the luteinizing hormone LH).

2. The LHRH stimulates the pituitary gland to release the luteinizing hormone LH and the hormone FSH (follicle-stimulating hormone).

3. The FSH acts on the Sertoli cells, which release the androgen-binding protein.

4. The LH stimulates the secretion of testosterone by the Leydig cells of the testicle.

5. The combination of testosterone and androgen-binding protein stimulates spermatogenesis. The testosterone circulating in the bloodstream results in the development of secondary sexual characteristics (development of male body hair, muscles, external genitalia, etc.).

6. The increased concentrations of testosterone and inhibin exercise a negative feedback effect on the hypothalamus and the pituitary gland, reducing the secretion of their hormones.

The systems of the human body
The reproductive system

Female genital organs

The female sex organs can be divided into external or visible organs, like the vulva, and internal organs such as the vagina, the uterus, the Fallopian tubes, and the ovaries.

The female external genitals, collectively known as the vulva, surround and constitute the external opening of the vagina. The vulva consists of the labia, the interlabial space and the clitoris.

The female internal genitals are the vagina, the uterus, the Fallopian tubes and the ovaries.

The vagina is a tubular organ extending from the vulva to the uterus. It is an important region for sexual sensations, and forms part of the birth canal.

The uterus is a fibromuscular organ, in two parts: the body of the uterus, and a lower cylindrical section communicating with the vagina known as the cervix. This plays an essential role in the process of fertilisation.

Ovum

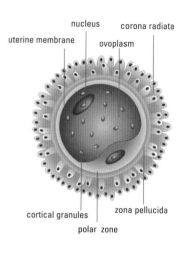

The two Fallopian tubes or oviducts are each about 12 cm long, lying on either side of the uterus and extending from each ovary to the apices of the uterus. They connect the uterine cavity with the abdominal cavity. Through these ducts pass the ovum and the spermatozoa, and eventually the fertilized ovum that will implant in the uterus.

The two ovaries lie one on each side of the pelvic cavity, attached by ligaments, though retaining a certain degree of mobility. Their primary function is the formation of ova and the secretion of hormones.

Anatomy

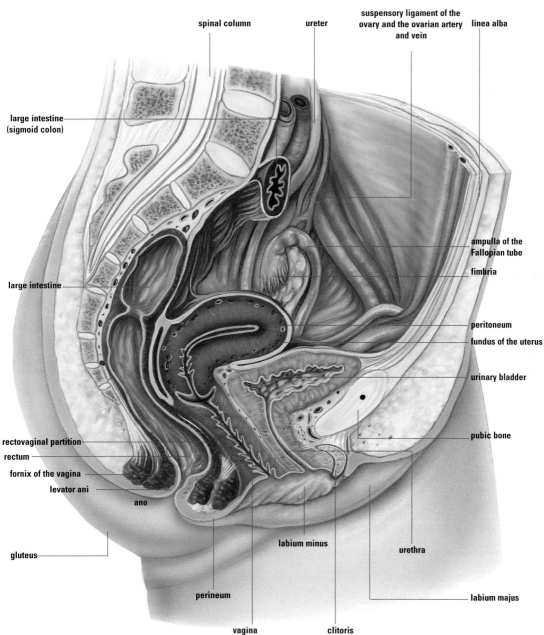

vagina
A flattened cylindrical fibromuscular tube extending from the vulva to the uterus, lying between the urinary bladder in the front and the rectum to the rear. At its upper end it joins the cervix, its mucosal lining widening and forming folds to produce the fornices of the vaginal sac. The vagina is the organ of copulation in females, and in sexual intercourse receives the male penis.

clitoris
A small erectile organ of great sexual sensitivity, situated in the frontal area of the vulva, the principal function of which is to achieve sexual stimulation leading to orgasm. Like the male penis, it contains spongy, erectile tissue and also many nerve endings.

The systems of the human body
The reproductive system

Function of the female genital organs

The menstrual cycle is controlled by hormones, specifically by the interaction of two hormones from the pituitary gland (follicle-stimulating hormone, FSH, and luteinizing hormone, LH), and the female sex hormones (estrogens and progesterone). These hormones are involved in the development of the ovum in the ovary and its expulsion, the growth of the endometrium in the uterus and its expulsion if not fertilized, which produces menstruation. The first day of menstruation corresponds to the first day of the menstrual cycle.

The development and maturation of an ovarian follicle, which contains the ovum, takes place under the influence of the hormones FSH and LH. During its growth in the first two weeks of the cycle this follicle produces increasing amounts of estrogen. Estrogen levels reach their maximum just before ovulation; this stimulates the maturation of the ovum and its expulsion, normally halfway through the cycle, around day 14.

The active hormone during the second half of the cycle is progesterone, secreted by the remains of the follicle after the ovum has left it (it is then known as the corpus luteum).

If the ovum is not fertilized the corpus luteum atrophies. The resulting decrease in progesterone and estrogen levels deprives the endometrium (the inner layer of the wall of the uterus) of its hormonal stimulation. The immediate result of this is the constriction of the uterine blood vessels, with a consequent fall in the amounts of oxygen and nutrients reaching the tissue. Without these, the endometrial tissue disintegrates and detaches itself from the uterus, resulting in the start of the menstrual flow. This marks the first day of the cycle, which normally lasts for about 28 days.

This pattern is interrupted if the ovum is fertilized by a spermatozoon. In this case the hormone levels are maintained by parts of the fertilized egg until the new placenta is formed and takes over hormone secretion.

Hormones are thus influential in the development of the menstrual cycle. Estrogen also produces changes in the characteristics of the external and internal genital organs, such as development during puberty of the labia minora and the breasts, and the growth of the uterus.

The menstrual cycle normally lasts about 28 days. The start of a period marks the first day of the cycle, with ovulation occurring on about the 14th day.

Function

The influence of hormones on female genital function

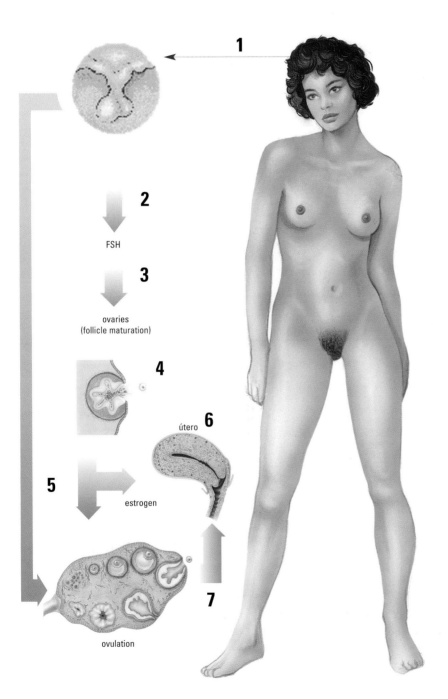

1. The levels of estrogen in the blood stimulate or impede the secretion of FSH and LH.

2. FSH and LH are produced in the pituitary gland.

3. FSH stimulates the maturation of ovarian follicles.

4. The maturation of the follicles, together with maximum secretion of LH, leads to ovulation.

5. The maturation of the follicles increases the secretion of estrogen.

6. The secretion of estrogen stimulates the growth of the endometrium in the uterus.

7. After ovulation, if pregnancy does not occur, the sudden drop in estrogen produces a period.

The systems of the human body

The nervous system

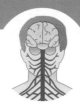

The nervous system is the most important control system in the human body. The other system involved in organ control is the endocrine system, but whereas this uses hormones as messengers, the nervous system uses electrical stimuli which travel a great deal faster.

The nervous system consists of the central nervous system and the peripheral nervous system.

The central nervous system

This consists of the brain and the spinal cord. It is in the brain that the "higher" senses, both cognitive and emotional, are found. It is also responsible for producing sensations and controlling movement.

The peripheral nervous system

The peripheral nervous system consists of all the nervous tissue outside the central nervous system: the peripheral nerves that innervate muscles and organs.

Nervous tissue consists of an intricate, interconnected network of specialized cells called neurons, which are enclosed within a supportive tissue that has the same function in the nervous system that connective tissue fulfils elsewhere. The characteristic support cells of the brain are known generically as glia, and include astrocytes and oligodendrocytes.

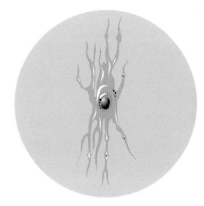

Fig. 1

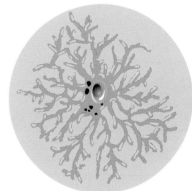

Fig. 2

Support cells of the central nervous system: an oligodendrocyte (Fig. 1) and an astrocyte (Fig. 2).

Anatomy

Organization of the nervous system

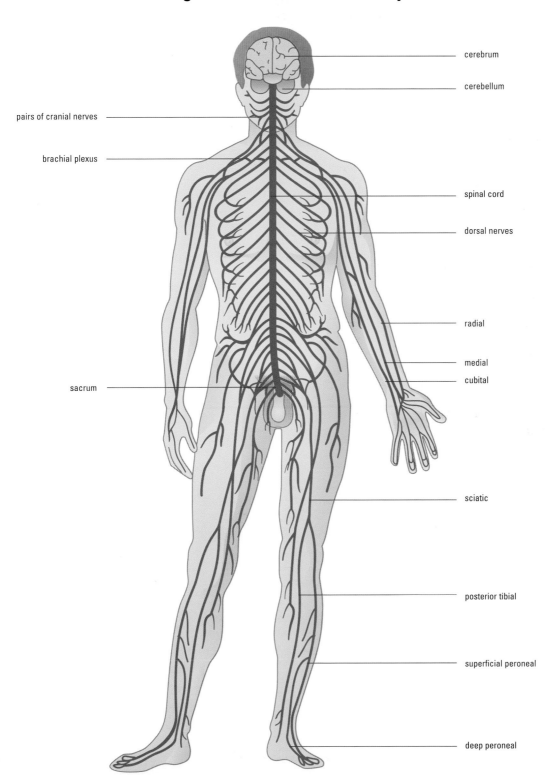

The systems of the human body
The nervous system

Neurons

The functional and structural unit of the nervous system is the neuron. Although neurons vary widely in form and size they all share the same basic structure, consisting of a cell body, which contains the nucleus surrounded by cytoplasm; an axon, an extension from the cell body, of variable length ending in a cluster of small protuberances or dendrites, multi-branched extensions of the cell body that make contact with other neurons.

The dendrites at the tips of the axons establish communication with other neurons through synapses (see next page). There is no direct contact between cells at a synapse; the neurons are separated by a synaptic cleft, across which stimuli are transmitted via chemicals called neurotransmitters.

Types of neuron

Unipolar

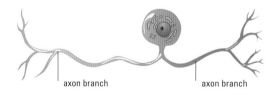

axon branch — axon branch

Bipolar

axon — axon

Multipolar

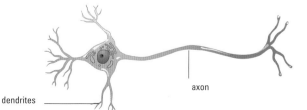

dendrites — axon

There are three basic types of neuron defined according to their form: unipolar, in which a single axon divides into two branches; bipolar, in which two axons originate from different points of the neuron cell body; and multipolar, in which an axon and many dendrites originate from the cell body.

Function

Synapse

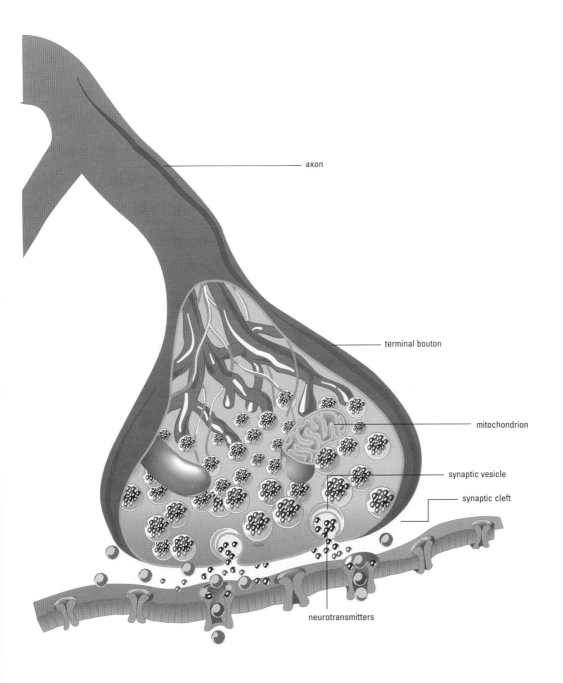

The nerve impulse passes down the axon of the first neuron to the terminal bouton, where the neurotransmitter molecules are released and pass across the synaptic cleft to the second neuron, carrying the nerve stimulus.

The systems of the human body

The nervous system

The nerves

The nerves are the elements responsible for conveying nerve stimuli in the peripheral nervous system. They form bundles and some are long, extending from the spinal cord to the tip of a finger or toe.

There are two types of nerve, defined according to function: the somatic nerves, which are involved in voluntary functions and are the type that stimulate the muscles to produce movement; and the autonomic nerves, which control involuntary functions such as the functioning of the different organs. For example, the heart is innervated by the autonomic nervous system, which increases or reduces the heart rate according to different circumstances (exercise, emotions, etc).

One of the most important nerves of the autonomic nervous system is the vagus nerve (named from the Latin, *vagare*, to wander), which controls many vital functions such as heart rate, digestion, and breathing.

Each nerve is formed by one or more bundles of nerve fibers. Each individual nerve fiber is the axon of a neuron which is covered by the cytoplasm of a supporting cell known as a Schwann cell.

Structure of the nerves

Nerves are structures of different thickness and length. The cell bodies of the neuronal axons that form the nerves are situated in the central nervous system or in the collections of cell bodies (ganglia) that lie next to the spinal cord.

Each nerve is formed by one or more bundles of nerve fibers. Each individual nerve fiber is the axon of a neuron which is covered by the cytoplasm of a supporting cell known as a Schwann cell. Large-diameter fibers are covered by several concentric layers of Schwann cells, which form a sheath of myelin. In addition to its supporting function, myelin is also responsible for increasing the speed of nerve impulses.

Each bundle of nerve fibers is surrounded by a layer of connective tissue called the perineurium; if the nerve contains many bundles, these are surrounded by another layer known as the epineurium.

Structure

Structure of the nerve

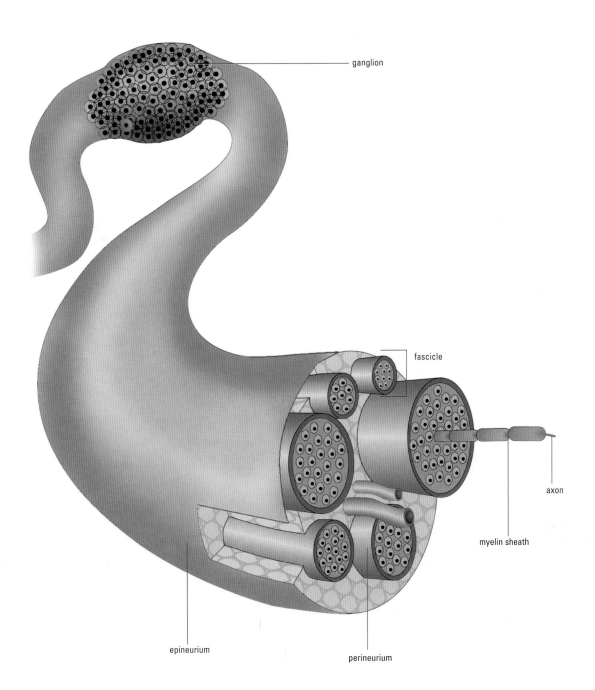

The systems of the human body
The nervous system

The brain

The cerebrum is the central organ of the nervous system. The cerebrum, the cerebellum, and the brain stem together make up the brain, which in an adult consists of more than 12,000 million neurons and weighs about 1.5 kg.

The outstanding anatomical features of the cerebrum are the cerebral convolutions—very irregular folds of cerebral tissue, the pattern of which varies from one person to another. These convolutions are separated by a series of furrows known as sulci. Some of these are very deep and are known as fissures. The two parts of the cerebrum, the cerebral hemispheres, are separated by the longitudinal fissure.

The outer layer of the cerebrum is the cortex (or cerebral cortex). It is formed by neuron cell bodies and because of its color is called the gray matter. It is in the cortex that the areas controlling voluntary movement, the senses, language, and vision are found. There are also areas relating to memory and thought.

The inner part of the central nervous system (CNS) is composed of white matter, and consists of axons covered in myelin.

It is also in the cerebrum that the most important nerves of the peripheral nervous system originate. There are twelve pairs of these cranial nerves (see page 131).

Cerebral cortex

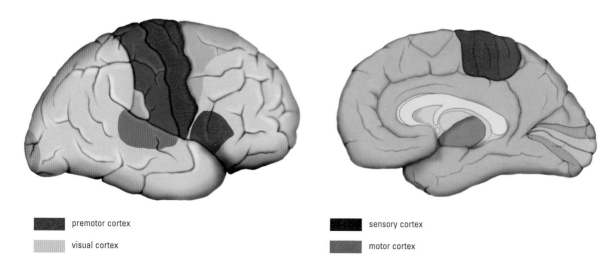

- premotor cortex
- visual cortex
- sensory cortex
- motor cortex

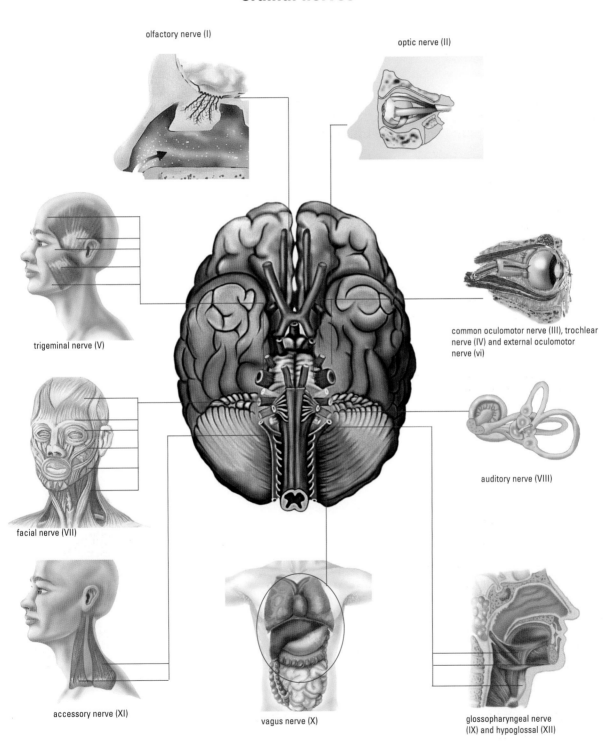

The systems of the human body
The nervous system

The spinal cord

The spinal cord is housed in the spinal column and passes through the canals of the vertebrae. It is approximately 45 cm long and extends from the brain stem to the lumbar sacral region of the lower back.

Thirty-one pairs of peripheral spinal nerves originate in the spinal cord, connecting it with the rest of the organism. These spinal nerves are formed in the space between two vertebrae by the fusion of two roots arising directly from the cord: a motor root and a sensory root. The motor root innervates muscles to produce movement, whereas the sensory root gathers information from the peripheral receptors and transmits it to the brain.

In the spinal cord, the relationship between gray matter and white matter is the reverse of that in the brain, i.e. the gray matter lies inside the white matter.

Inside the white matter groups of nerve fibers follow different tracks. Some come from the cerebrum (descending fibers) while others come from the periphery and carry stimuli to the cerebrum (ascending fibers).

The spinal cord would be very vulnerable if it were not protected by the bony segments of the spinal column and supporting ligaments. In fact lesions to the spinal cord produce paralysis below the site of the lesion as nerve transmission is broken from that point onward.

The cerebrum is covered by three layers of connective tissue, the meninges: from outer to inner they are the dura mater, the arachnoid, and the pia mater. These also cover the spinal cord, thus providing another element of protection.

The pia mater is the meninx most closely related to the spinal cord, and lies closely against it. Between the dura mater and the arachnoid meninx is the subarachnoid space, filled with cerebrospinal fluid.

Structure

The spinal cord

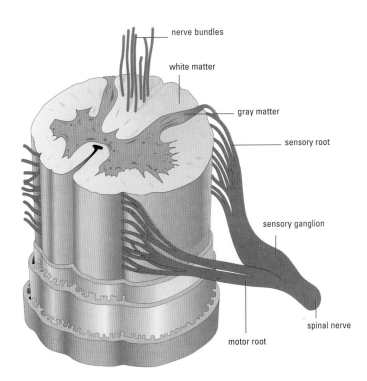

The autonomic nervous system

The autonomic nervous system controls the involuntary activity of most organs and glands.

The autonomic nervous system regulates the internal functions of the organism, maintaining physiological equilibrium (homeostasis). It controls the involuntary activity of most organs and glands. It is therefore fundamental to such processes as digestion, heart beat, and hormone secretion.

The part of the cerebrum that processes information from the autonomic nervous system is the hypothalamus. From here information passes to the spinal cord through the brain stem.

The autonomic nervous system subdivides into the sympathetic and parasympathetic systems (see pages 134 and 135). The sympathetic system is an excitatory system that prepares the organism for situations of both physical and psychological stress. The parasympathetic system maintains and restores energy.

The systems of the human body
The nervous system

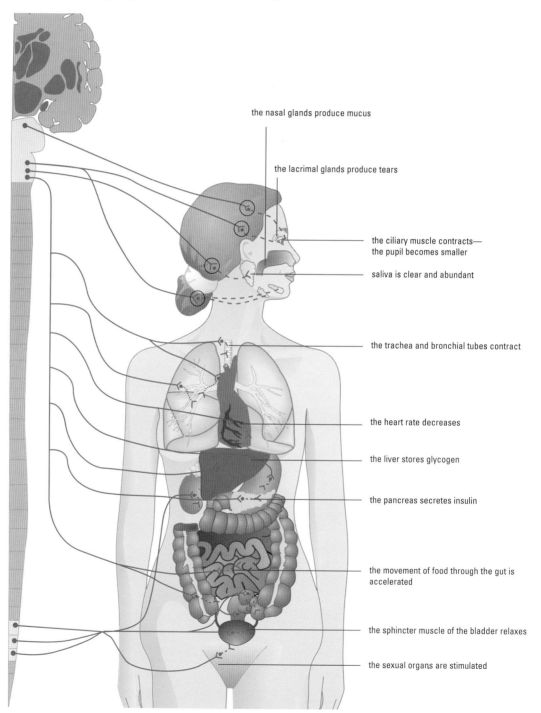

The parasympathetic nervous system: effects of stimulation

- the nasal glands produce mucus
- the lacrimal glands produce tears
- the ciliary muscle contracts—the pupil becomes smaller
- saliva is clear and abundant
- the trachea and bronchial tubes contract
- the heart rate decreases
- the liver stores glycogen
- the pancreas secretes insulin
- the movement of food through the gut is accelerated
- the sphincter muscle of the bladder relaxes
- the sexual organs are stimulated

Structure

The sympathetic nervous system: effects of stimulation

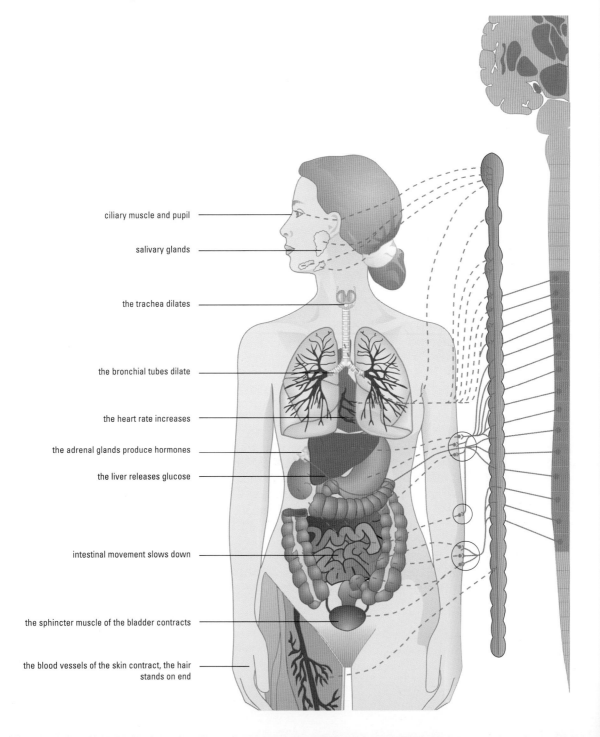

The systems of the human body

The endocrine system

The endocrine glands are the organs in which hormones are produced and from which they are released into the blood. When each hormone reaches its target organ (the organ that has receptors for it and responds to its action), it either inhibits or stimulates the activity of that organ. The various glands are interrelated, and some control the activities of others through a series of feedback mechanisms in which excess of one hormone results in the inhibition of production of another that stimulates production in the first. The nervous system and the endocrine system together control the activities of all the organs and systems of the human body.

The principal endocrine glands are the pituitary gland and the pineal gland (situated in the cranial cavity), the thyroid and parathyroid glands (in the neck), the adrenal glands (above each kidney), the pancreas and the ovaries (in the abdominal cavity), the testes (in the scrotum), and in pregnant women the placenta. Each is responsible for the production of specific hormones, such as the hormone that controls growth (growth hormone), the hormone that enables a cell to take up glucose (insulin), those that produce the sexual characteristics of each sex (androgens and estrogens), the hormone that stimulates the female mammary gland to secrete milk (prolactin), and so on. In this way, the endocrine system controls such important and widely different processes as growth, nutrition, and reproduction. The endocrine glands, the hormones they produce, and their target organs thus together constitute a major system of regulation and control of the entire organism.

Although endocrine glands possess some characteristics that differ widely, they all have certain elements in common. All possess specialized secretory cells surrounded by supporting cells and abundant vascular tissue, into which the synthesized hormone passes. Certain glands, such as the thyroid and the adrenal glands, form organs; others are associated with organs such as the pancreas or the kidney.

Unlike the effects of nervous stimuli, which appear rapidly but may be short-lived, the effects of hormones appear more slowly, last for longer, and are usually apparent in tissues that are quite remote from the gland where they originate.

Hormones regulate such varied processes as growth, the control of blood pressure and the response of the organism to attack.

Anatomy

Principal endocrine glands

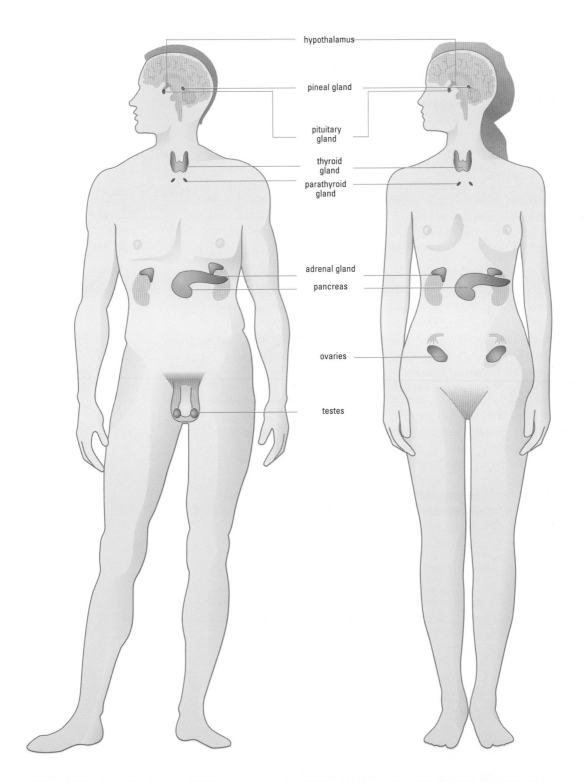

The systems of the human body

The endocrine system

The principal glands of the human body

The thyroid and parathyroid glands are found in the neck, in front of the trachea. The thyroid produces the hormones T3 (tri-iodothyronine) and T4 (tetraiodothyronine or thyroxine). Thyroid hormones control basal metabolism and are important in the maturation of the nerve tissue. Their production is under the control of the pituitary gland through the manufacture of TSH (thyroid-stimulating hormone, or thyrotropin). The thyroid stores large quantities of the thyroidal hormones, which are produced in globular sacs known as the thyroid follicles. Around these follicles are groups of clear cells called parafollicular cells; these are stimulated by a rise in the level of calcium in the blood to secrete the hormone calcitonin, which incorporates this excess calcium into bone tissue.

The parathyroid glands are four small structures on the posterior surface of the thyroid gland. They have a role in the control of calcium levels through the production of the parathyroid hormone. This hormone raises the levels of calcium in the blood and, together with calcitonin, controls the concentration of calcium in the blood.

The adrenal glands are two small glands situated above the kidneys, each consisting of an outer region called the cortex and an inner medulla. The cortex produces mineralocorticoids, glucocorticoids, and sex hormones. The mineralocorticoids are involved in maintaining the balance of fluids and ions in the body. The glucocorticoids play an important role in the metabolism of fats, proteins, and carbohydrates. The small quantity of adrenal sex hormones complements the other sex hormones. The adrenal medulla produces adrenalin and noradrenalin, both of which are important in adapting the organism to cope with stress situations.

The pancreas contains groups of cells specialized in manufacturing and secreting hormones, called the islets of Langerhans. Each islet contains alpha cells that produce the hormone glucagon, beta cells that secrete insulin, and delta cells that control other pancreatic hormones. Insulin and glucagon are both protein hormones. Insulin stimulates the uptake of glucose by the cells, thus reducing the amount of glucose in the blood after a meal. The metabolic effect of glucagon is almost the reverse. The secretion of both these hormones is controlled by the concentrations of glucose in the blood.

The thyroid hormones control basal metabolism and are important in the maturation of the nerve tissue.

Examples

Examples of glands

Thyroid and parathyroid glands (posterior view)

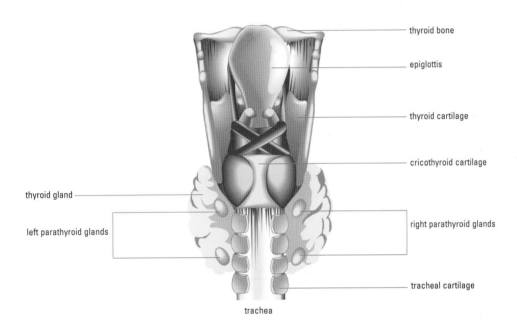

Adrenal gland

Pancreas

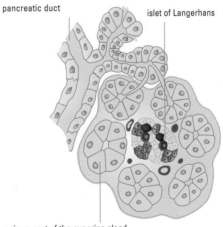

The systems of the human body
The endocrine system

Sex hormones

The production of the hormones that govern the sexual lives of men and women does not begin until puberty. In men the testes produce the male sex hormones (androgens) such as testosterone; in women the ovaries produce estrogen and progesterone. The tasks of these sex hormones are, on the one hand, the control of the production of sperm and ova and, on the other, the development of secondary sex characteristics, such as facial hair in men and the growth of breasts and the onset of menstruation in women.

Sex hormones are chemicals derived from steroids produced in the adrenal gland. Unlike protein hormones, which bind to specific receptors on the surface of the target cell, steroid hormones must enter the cell to interact with its receptors.

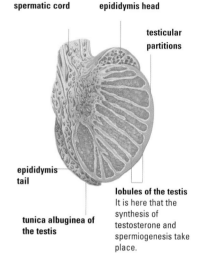

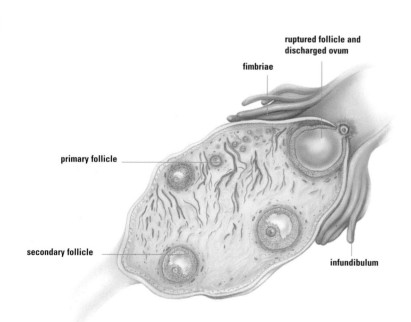

Top right: sagittal section of the testis, showing the interior view; note the partition of the lobules by septa.

Above: ovulation: the ovum is discharged into the abdominal orifice of the Fallopian tube and from there is carried to the ampulla.

Mechanism

The mechanism of hormonal control

The production of most hormones is controlled by feedback mechanisms. The hypothalamus, the pituitary gland, and the target gland are all involved in this kind of process. Feedback control allows a hormone to be released when it is needed; an excess of a hormone can inhibit its own production (negative feedback). This involuntary control mechanism maintains the balanced functioning of the body.

The diagram below illustrates the female sex hormone control mechanism: the synthesis of estrogen in the target gland (the ovary) is stimulated by the pituitary hormones FSH and LH. The pituitary gland is controlled by the hypothalamus of the brain. High estrogen levels in the blood have a negative (inhibitory) effect on hormonal synthesis in the hypothalamus and the pituitary.

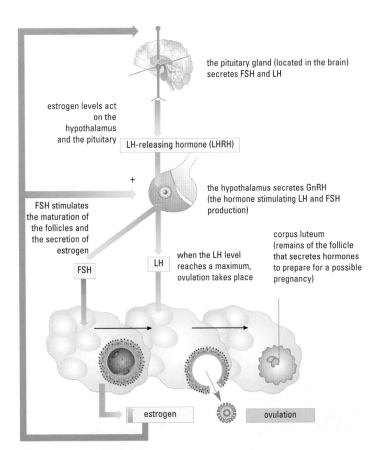

Right: Hormonal regulation of the menstrual cycle.

The systems of the human body
The endocrine system

The pituitary gland

The pituitary gland (also known as the hypophysis, or sometimes as the master gland) is one of the most important glands of the human body: it controls the operation of several vital organs and bodily functions.

Structurally no larger than a bean, it is suspended from the hypothalamus, to which it is connected by a short stem. It consists of two lobes, the anterior lobe or adenohypophysis and the posterior lobe or neurohypophysis. It produces two types of hormone. Some, like the growth hormone, affect their target tissue directly, while others, like the thyroid-stimulating hormone TSH, act to regulate hormonal production by another gland in the organism. Its functions are regulated in both cases by the hypothalamus, either through nerve impulses or hormonally.

The pituitary or master gland controls the development and functioning of important organs and systems of the body.

Pituitary hormones

The neurohypophysis manufactures two hormones: antidiuretic hormone (ADH or vasopressin), which controls the amount of water reabsorbed in the kidney, and oxytocin, which is concerned in the mechanisms of birth and lactation. These substances are produced in neurosecretory cells in the hypothalamus, and pass into the neurohypophysis through the axons of these cells. When they are needed they are released.

The adenohypophysis produces trophic hormones (which regulate the production of other glands) such as the thyroid-stimulating hormone TSH, adrenocorticotropin hormone ACTH, follicle-stimulating hormone FSH and luteinizing hormone LH. TSH controls the production of thyroid hormones in the thyroid gland. ACTH controls the production of hormones such as adrenalin and steroid hormones, in the adrenal gland. FSH and LH are gonadotrophic hormones that regulate growth and function of the sex organs.

The adenohypophysis also produces hormones that act directly on their target tissues. These include growth hormones and also prolactin, which stimulates lactation following birth and plays a part in the control of sex hormones. The functions of the adenohypophysis are in turn regulated by hormones produced by the hypothalamus, which reach the pituitary gland via the blood vessels connecting the two.

Anatomy

Pituitary gland

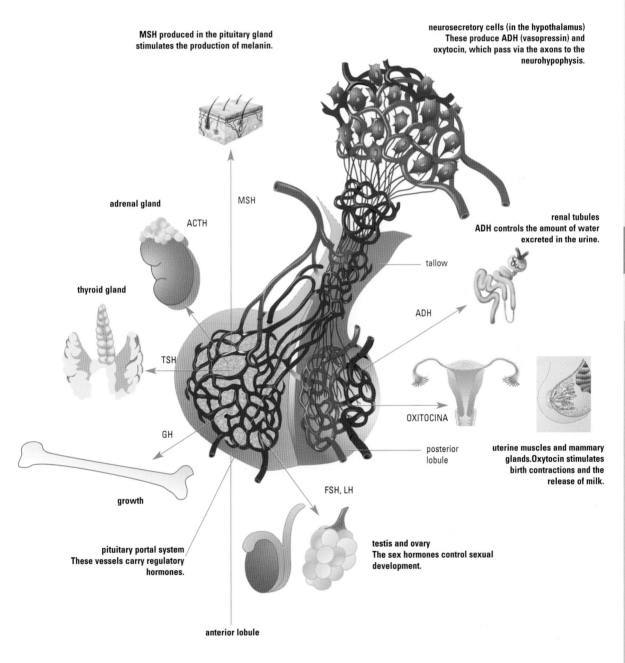

The systems of the human body

The immune system

All higher living things are equipped with a system that enables them to identify any substance foreign to the organism and to defend the organism against it. This is the immune system.

This defense system starts at the skin and the mucosa, which constitute the first barrier against micro-organisms. In addition to acting as a physical barrier, these surfaces also produce antimicrobial secretions, such as the lysozyme found in saliva and tears.

If a foreign substance passes through the skin and comes into contact with the internal tissues an immune response is unleashed. Through this response the organism endeavors to neutralize the foreign body. The cells that partake in this immunological response are macrophages such as B and T lymphocytes.

Once in the tissues, the invading agent is picked up by the lymph and transported to the lymphatic system, which is composed of numerous interconnected groups of lymphatic nodes. Each group of nodes drains a specific part of the body. Thus when an infection occurs, it is the lymphatic nodes closest to the site of the infection that become inflamed. The lymphatic system may also be used by cancer cells to produce remote metastases.

Other organs belonging to the immune system are the thymus, the spleen, and the tonsils.

The **thymus** is a lymphoid organ that lies in the lower part of the neck; it reaches maximum development in childhood and almost disappears in adulthood. Its function is to use the lymphocytes formed in the bone marrow to produce lymphocytes ready to respond to an immunological need: thus it is an organ where lymphocytes mature. The function of the spleen was described in the chapter on blood (see page 33).

Once in the tissue, the invading agent is picked up by the lymph and transported towards the lymphatic system, which is composed of numerous interconnected groups of lymphatic nodes.

Anatomy

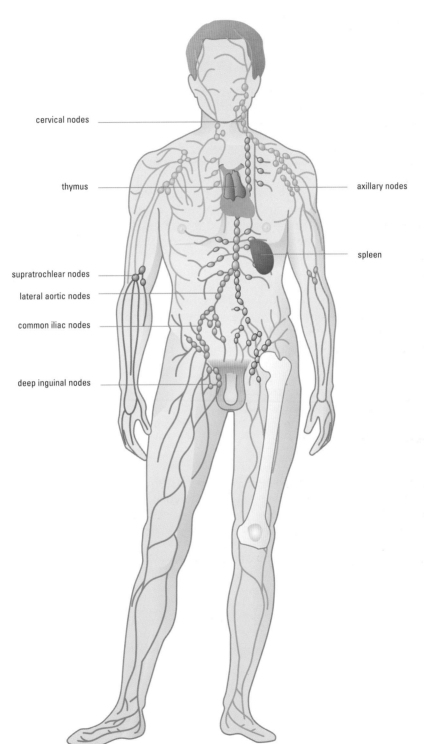

The systems of the human body
The immune system

The lymph nodes

The glands known as lymph nodes are distributed throughout the body. Each consists of a capsule of connective tissue containing immunological cells, mainly macrophages and lymphocytes. The size of the nodes depends on the part of the body in which they are found and on their state of activation. The axillary nodes, for example, may be the size of a lentil while the inguinal nodes may be up to 1.5 cm across in normal circumstances, but when a node is immunologically activated its size increases considerably, due to the increase in lymphocytes.

The **lymph nodes** are normally bean-shaped. Strips of connective tissue from the capsule form trabeculae of different lengths within the lymphoid tissue. Lymph arrives at the node from the afferent vessels via a space between the capsule and the lymphoid tissue known as the subcapsular sinus. It then passes through a system of channels, the medullary cords, to the hilus and leaves via the efferent lymph vessel.

The **lymphoid tissue** of the lymph node contains two distinct regions: the central medulla, and the cortex which surrounds it. In the cortex, lymphocytes are formed in lymph follicles, some of which have clear centers known as germinal centers.

The lymph nodes are supplied with blood through arteries, which enter the node through the hilus and form branches extending through the medulla, producing a network of fine capillaries distributed through the cortex and the medullary cords.

Within the lymph nodes, lymphocytes are distributed in such a way that the superficial cortex and the medulla contain mainly B lymphocytes while the central cortex contains T lymphocytes. Macrophages are found in the subcapsular sinus.

The number and size of lymph follicles depends on the state of activation of the node and the type of immune response. A cell-mediated response (via lymphocytes) produces paracortical enlargement, while a humoral response (via antibodies) produces a large number of cortical follicles with germinal centers.

The lymphoid tissue of the lymph node contains two distinct regions: the central medulla, which is surrounded by the cortex. Within the cortex, lymphocytes are formed in the lymph follicles.

Anatomy

Section of a lymph node

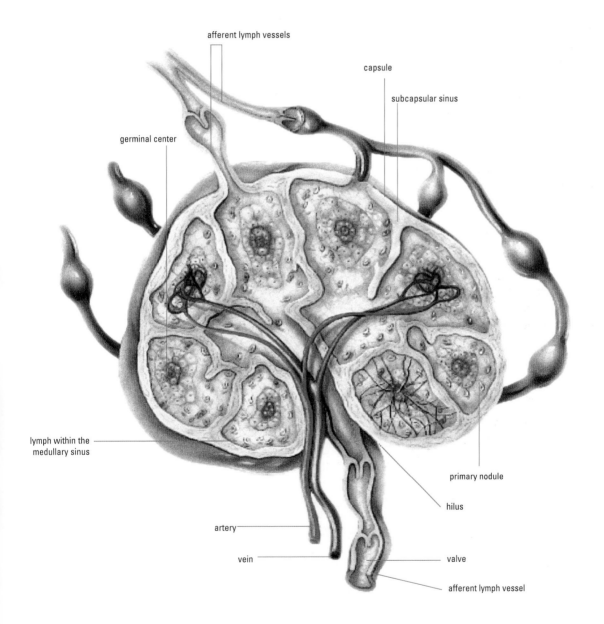

The systems of the human body
The immune system

Immune response

The immune response is the response of the cells of the immune system to invasion by foreign substances, which are known as antigens. The immune response is mediated by lymphocytes.

Lymphocytes circulating in the body can detect specific antigens, but require modification to determine the type of response that is needed.

Cell-mediated immune response

This is mediated by T lymphocytes which, after a phase of preparation in the thymus, enter circulation and spread throughout the lymphoid tissue of the organism. When a T lymphocyte encounters an antigen in a tissue it identifies it and then returns to the local lymph tissue where it "informs" other lymphocytes. These are then transformed into activated T lymphocytes, which will combat the antigen by means of two mechanisms: either by the manufacture of proteins called lymphokines or by direct destruction.

Humoral immune response

This is mediated by B lymphocytes. These are situated in the lymph organs, principally in the nodes and the spleen, where they lie in wait for antigens. When they encounter an antigen they are transformed into larger plasma cells, which secrete antibodies that will bind to the antigen and destroy it. Some of these cells, called memory cells, remain in the lymphatic nodes; if this antigen invades the body again the memory cells begin at once to secrete the appropriate antibody.

The mechanisms described above constitute what is known as specific immune response, which requires prior processing of the antigen. There is also the non-specific immune response, which consists of the direct and spontaneous destruction of antigens, including certain bacteria, by phagocytic cells in the connective tissue. These cells are primarily macrophages and neutrophil-type white blood cells.

In some types of cell, infections produced by viruses cause the secretion of antiviral substances known as interferons, which prevent the multiplication of viruses within the cell itself.

Lymphocytes circulating in the body can detect specific antigens, but require modification to determine whether a cell-mediated or humoral response is required.

Response

Another element of the non-specific immune response is the inflammatory response, described on page 150, which involves local circulatory changes and the attraction of phagocyte cells to the site of infection.

In addition, a group of some 25 blood proteins called the complement system plays an important part in the immune response. These circulate in the blood and are activated by antibodies (immunoglobulins). When an antibody molecule attaches itself to a foreign cell it activates complement proteins. The result is a cascade of successive reactions as they join together until they eventually form a pore in the surface membrane of the invading agent. This pore irretrievably damages the permeability of the membrane, and the micro-organism is destroyed.

Humoral immune response

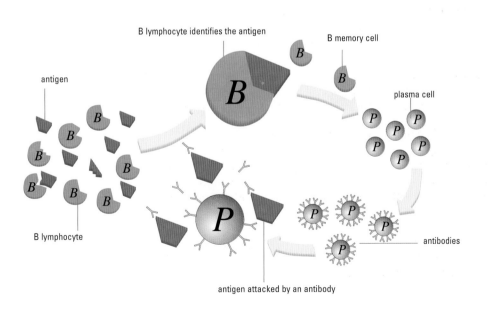

The systems of the human body
The immune system

The inflammatory response

The inflammatory response is one of the first mechanisms of defense to act when an antigen penetrates the skin or the mucosa.

It forms part of the non-specific immune response, attacking any micro-organism of the same form. Firstly, local blood flow increases to carry phagocytic cells (macrophages and neutrophil cells) to the affected area. Inflammatory substances such as histamine and prostaglandins are secreted, which increases the permeability of the capillaries.

The phagocytes attach themselves to the invader by means of the specific antibodies that react to it. This is followed by phagocytosis, the process by which the phagocyte surrounds and engulfs the antigen, after which it is digested and destroyed (see page 151).

The inflammatory response can be stimulated not only by infectious agents such as viruses and bacteria, but also by physical or chemical attack. For example, the continued scratching of an area of skin produces an inflammatory response even when there is no infection. Burns can produce a more or less intense inflammatory response, as can a direct blow to any part of the body.

Occasionally an inflammatory response occurs without any external stimulus, as in allergic reactions. In such cases a specific antigen may trigger a disproportionate inflammatory response, which can range from reddened weals on the skin (urticaria) to intense inflammation of the bronchial tract and the consequential respiratory problems. An extreme example is anaphylaxis, when the inflammatory response affects the cardiovascular system, with excessive dilation of the blood vessels and a subsequent drop in blood pressure which can be fatal.

The inflammatory response can be stimulated not only by infectious agents such as viruses and bacteria, but also by physical or chemical attack.

Response

Inflammatory response

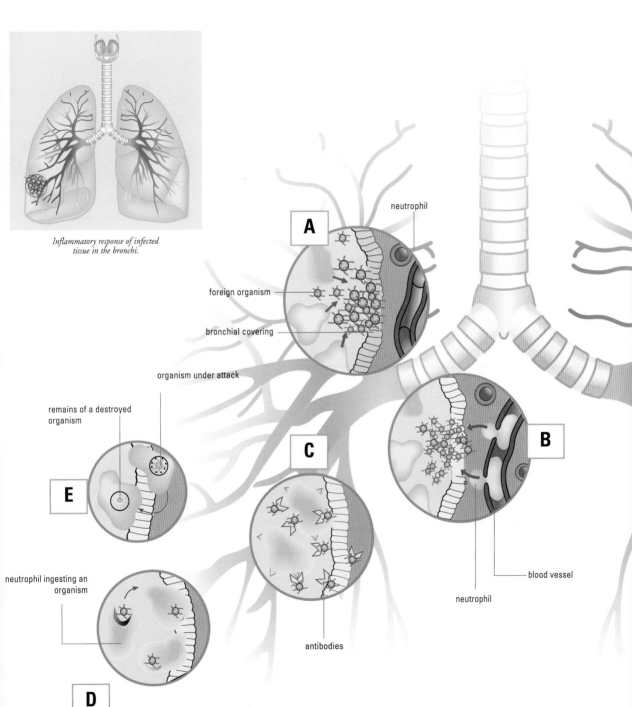

Inflammatory response of infected tissue in the bronchi.

The systems of the human body

The urinary system

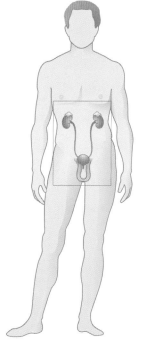

The principal task of the urinary system is to eliminate the waste products of metabolism, such as urea and creatine, from the body. The basic aim is to maintain the composition of the internal environment constant, in terms of both body fluids and the substances dissolved therein. The entry of excess water and/or salts into the organism is therefore compensated for by an adjustment in the proportion of these in the urine.

Other important functions of the urinary system include helping to maintain arterial pressure and stimulating the production of red blood cells in the bone marrow. Both are regulated by hormones.

Anatomy

The urinary system consists of two kidneys, lying against the dorsal wall of the abdomen to the left and right of the spinal column, two ureters that feed into the urinary bladder, and the urethra through which the body eliminates urine.

The kidneys are between 10 and 12 cm in length. Each consists of two regions, an outer cortex and an inner medulla. It contains around a million filter units known as nephrons. Blood is filtered in the kidneys to produce urine, which is eliminated via the renal pelvis, the ureters, the bladder, and the urethra. Lying above each kidney is one of the adrenal glands, which produces hormones.

The ureters are ducts that carry urine from the kidneys to the urinary bladder. In a normal adult they are about 30 cm long and have a diameter of 5 mm. From their starting point, the renal pelvis, they pass behind the peritoneum and enter the bladder at such an angle that the bladder musculature acts as a sphincter, preventing the backflow of urine.

Anatomy

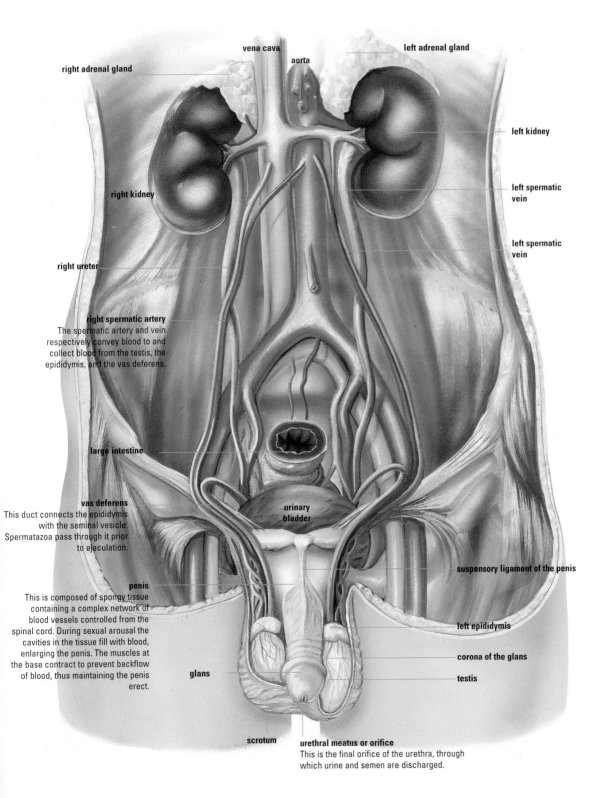

The systems of the human body

The urinary system

The urinary bladder lies in the pelvic cavity, close to the uterus in women and to the prostate gland and the seminal vesicles in men. Its muscular walls can dilate to store urine and contract to eliminate it.

The urethra is the duct connecting the urinary bladder with the exterior. It provides a passage for urine, and—in men—for semen as well. In males the urethra is 20 cm in length and consists of three sections: the spongy urethra, the membranous urethra, and the prostatic urethra. The female urethra is only about one-fifth as long as that of the male.

The prostate gland in a normal adult male weighs approximately 20 g. It consists of two lobes located in a lateral position around the neck of the bladder and the urethra, and a middle, smaller lobe. Its action is hormone-dependent (controlled by hormones), and its size normally increases with age.

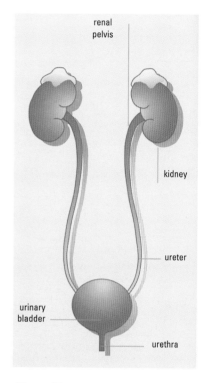

Diagram of the excretory system

Physiology

Blood carrying waste substances passes through the renal artery to the nephrons, the structural elements of the kidney. These are a series of thin-walled tubules in close contact with the arterial capillaries. An exchange of substances takes place between the fluid in these tubules and the capillaries, resulting in the formation of urine—a yellowish watery liquid, which on leaving the kidney passes to the bladder and is eliminated from the body.

The total volume of blood circulating in the body passes through the kidneys around 300 times a day. On average, about 1.5 liters of urine is excreted every day.

The total volume of blood circulating in the body is filtered in the kidneys around 300 times every day.

Anatomy

Anatomical relationship of the ureters in the female

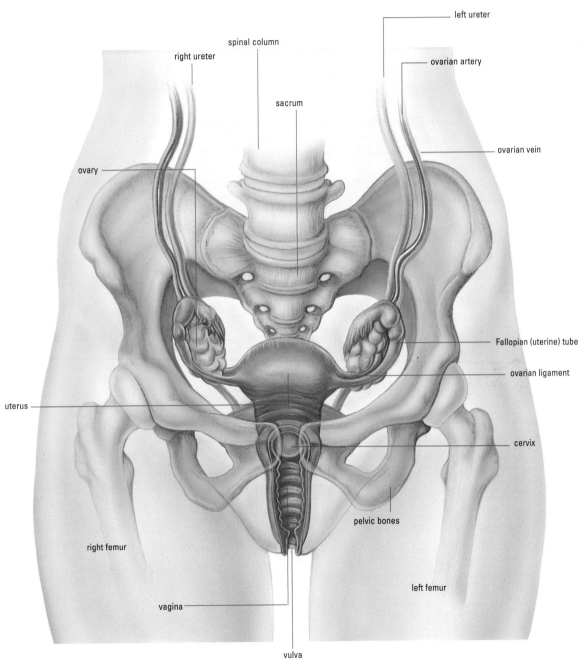

The systems of the human body
The urinary system

Vascularization of the kidneys

The main renal artery originates at the aorta, and divides into two main branches at the renal hilus. Each branch subdivides into several interlobar arterioles, which pass between the renal pyramids to the boundary between the renal cortex and the medulla. Here they divide again to form the arcuate arteries, then the interlobilar arteries and finally afferent arterioles leading into the glomerulus. Blood leaves the glomerulus via the efferent arteriole of the glomerulus and a series of venules that converge to form the renal veins. The ureters closely follow the paths of the major blood vessels that run through the posterior part of the abdomen, that is, the aorta, the renal arteries, the inferior vena cava, and the renal veins.

Veins of the abdominal cavity

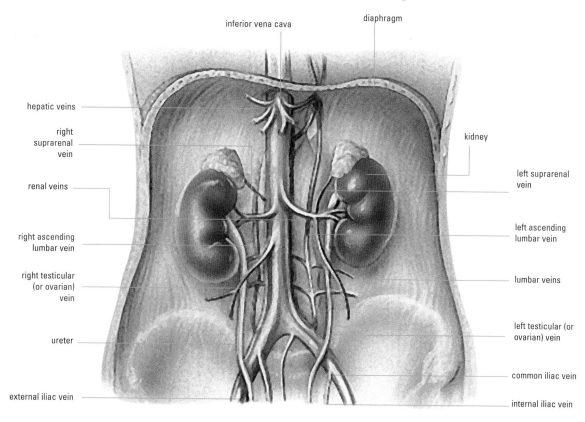

Structure

The intra-renal arteries

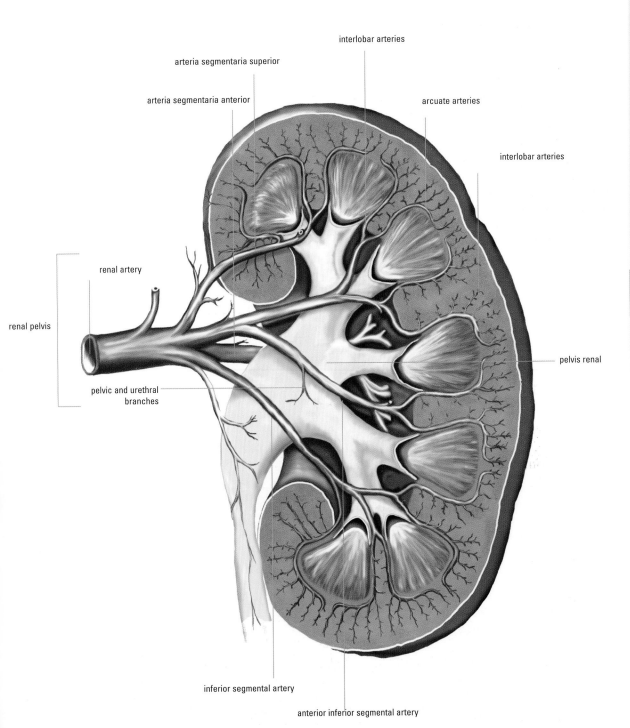

The systems of the human body

The urinary system

The nephron

Nephrons are the functional units of the kidney. Each kidney contains approximately one million nephrons. Their principal task is to filter the blood plasma passing through them and then to reabsorb part of the water and any useful substances, allowing the waste materials to be eliminated through the urine.

The nephron consists of two parts, the renal corpuscle and the renal tubule.

The renal corpuscle

This is the part of the nephron where plasma is filtered. The renal corpuscle is composed of a capsular structure, called Bowman's capsule, which surrounds the glomerulus, a tangled network of blood capillaries. Lying between the Bowman's capsule and the glomerulus is a layer of epithelial cells known as podocytes, which play an important role in glomerular filtration. The structures of the nephron and the renal corpuscle are shown on pages 159 and 161. The glomerulus starts at the afferent arteriole, which branches to form capillaries; at their venous end the capillaries converge and leave the glomerulus via the efferent arteriole. Glomerular filtration begins in a narrow space known as Bowman's space (between the podocytes and the capillaries); the filtrate then passes through the renal tubule system.

The renal tubule

The renal tubule is approximately 55 mm long in the male. Its function is to reabsorb part of the water and other substances that passed through the glomerulus. Different sections perform slightly different functions: the first or proximal convoluted tubule is the longest and reabsorbs most of the water and ions from the first glomerular filtration. Next is the loop of Henle: this is the part of the tubule that passes from the cortex deep into the renal medulla, creating significant changes of pressure. It continues into the distal convoluted tubule, the cells of which regulate sodium and potassium levels, and finally into the collecting tubule that carries urine to the bladder.

The osmotic pressure varies widely between the different regions of the kidney, providing a mechanism that balances the flow of water and dissolved substances.

Structure

Structure of the nephron

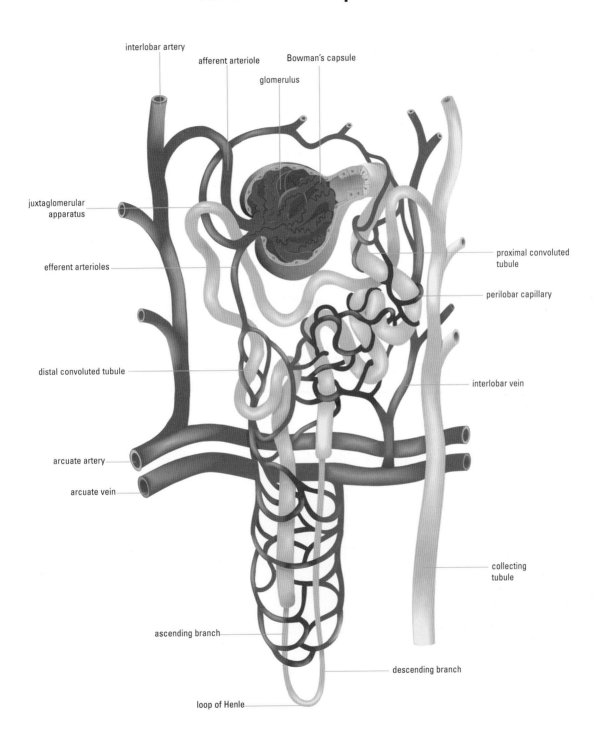

The systems of the human body
The urinary system

The juxtaglomerular apparatus

The glomerulus is closely associated with the juxtaglomerular apparatus, the function of which is to control arterial pressure. It is formed by the part of the afferent arteriole immediately before it enters the glomerulus and by an adjacent section of the distal convoluted tubule (see pages 159 and 161). The juxtaglomerular cells originate in the wall of the arteriole and produce the enzyme renin. This substance brings about an increase in the reabsorption of sodium ions and water in the distal convoluted tubule, which leads to a rise in arterial pressure. Any drop in pressure is detected in the arteriole of the glomerulus, which leads to the release of renin into the blood resulting in the restoration of the arterial pressure.

Within the juxtaglomerular apparatus are specialized cells that produce erythropoietin, a hormone that stimulates erythrocyte formation in the bone marrow.

As well as removing waste substances from the blood, the urinary system is involved in regulating blood pressure and stimulating the production of red blood cells in the bone marrow.

Formation of urine

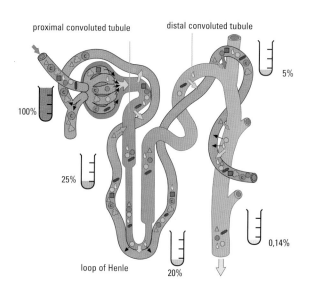

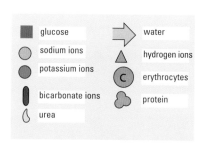

Function

Glomerular filtration

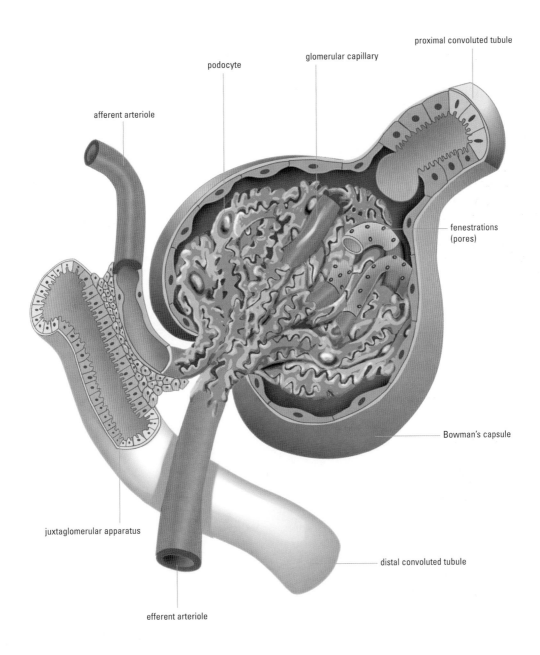

The systems of the human body

The senses

Sight

The eye is the organ of sight. It is a highly specialized organ, and is considered to be an extension of the brain. The phenomenon of vision consists of the conversion of light waves into nerve impulses transmitted to the brain, which processes this information and creates an image. This conversion takes place via special receptors which are extensions of the nerve cells found in the retina. These receptors are known as photoreceptors and are of two types: cones, which produce color vision, and rods which enables vision in shades of gray.

The eye is composed of three basic layers.

The sclera, the outermost layer, a hard fibroelastic layer which supports the eye. The external muscles of the eye are inserted into the sclera, which is opaque with the exception of the extreme anterior section; this is transparent and forms the cornea. The cornea is the eye's principal means of refraction and, together with the lens, is responsible for focusing images on the retina.

The choroid forms the middle layer. It is highly vascularized, and is pigmented to absorb the light that reaches the retina. In the anterior section the choroid connects to the ciliary body, which is a thickened area that is continuous with the iris. The ciliary body contains smooth muscle, the contraction of which regulates the shape of the crystalline lens. The lens itself is a true lens, the function of which is to focus images on the retina. Increased opacity of the lens is what produces cataracts. The crystalline lens and ciliary body together separate the eye into the anterior segment (which is in turn divided into anterior and posterior chambers) and the posterior segment. The anterior segment contains a liquid known as aqueous humor, while the posterior segment, to which the retina belongs, contains a gelatinous material called vitreous humor. The iris is pigmented and is the part of the eye that gives it its "color". Light enters the eye through the central part of the iris, the pupil; the iris acts as a diaphragm controlling the size of the pupil, and hence the amount of light entering the eye.

The retina is the innermost layer of the eyeball and is composed of nerve fibers containing photoreceptors. All the nerve fibers converge at a single point to form the optic nerve, which connects directly to the brain. The part of the retina aligned with the visual axis of the eye is known as the macula lutea (yellow spot), and at its center is the fovea, the area of sharpest vision.

Cataracts, which impair normal vision, are the result of a loss of transparency in the crystalline lens.

Structure

Structure of the eye

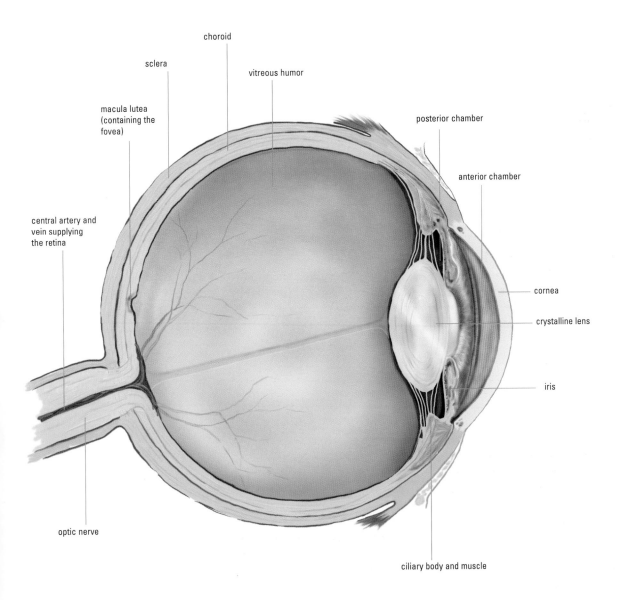

The systems of the human body
The senses

Hearing

The ear is the organ of hearing and balance. It is divided into three sections: the outer ear, which protects the middle ear and allows sound waves to pass into it; the middle ear, which transmits this movement to the inner ear; and the inner ear which translates these vibrations into nerve messages to be interpreted by the brain.

The outer ear consists of the pinna and the external auditory canal. The latter is covered in cilia and wax-secreting glands.

The middle ear is a cavity in the temporal bone between the tympanum and the inner ear. Three small bones known as the malleus, incus, and stapes (or, from their shapes, as the hammer, anvil, and stirrup) transmit vibrations to the inner ear.

The inner ear, also known as the labyrinth, consists of three semicircular canals, the vestibule, and the cochlea. The sense of hearing is located here, in the organ of Corti. The sense of balance is located in the structures of the vestibule and the semicircular canals.

The Eustachian tube connects the middle ear with the nasopharynx.

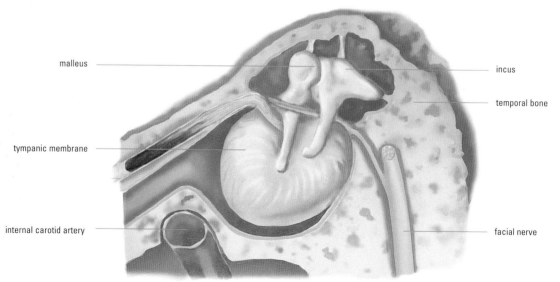

Tympanic cavity

Structure

Structure of the ear

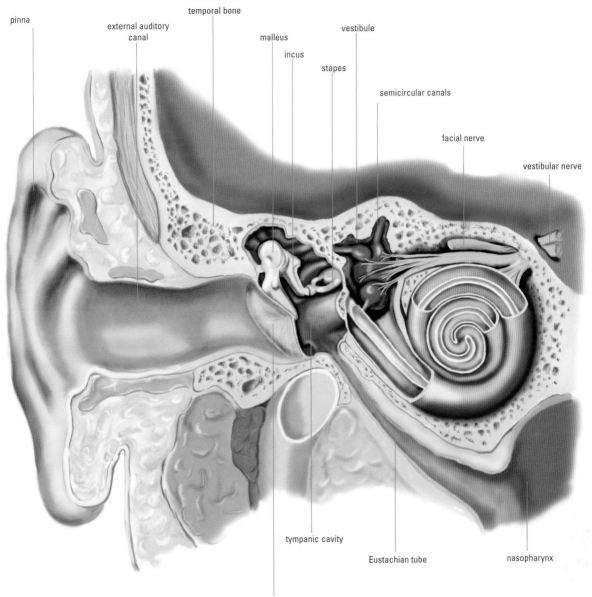

The systems of the human body

The senses

Smell

The receptors for the sense of smell are situated in a specialized part of the epithelium that lines the roof of the nasal cavity at the ethmoid bone, and is known as the olfactory epithelium. The olfactory cells of this epithelium are bipolar neurons, carrying at one end specialized cilia that capture the sensation of smell produced by odor molecules, diluted in the mucus produced by neighboring cells in this same epithelium. The axon at the other end of the olfactory cell converges with other axons from the area to meet in the olfactory nerve, which travels to the olfactory bulb (part of the brain).

The mechanism of smell

The odor molecules that enter the nose dissolve in nasal mucus and stimulate the nerve endings of the olfactory cells, generating nerve impulses that travel via the cell axons through the cribiform plate of the ethmoid bone to the olfactory bulb. From here the sensation of smell is transmitted to the brain via the olfactory nerves.

In humans the olfactory epithelium occupies a much smaller area than in other animal species.

The mechanism of smell

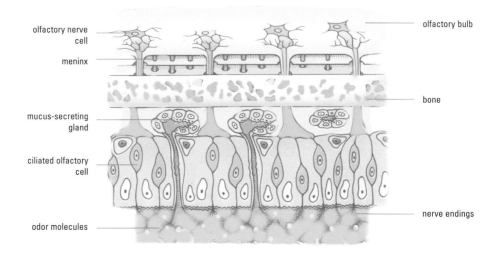

The nose

Olfactory structures

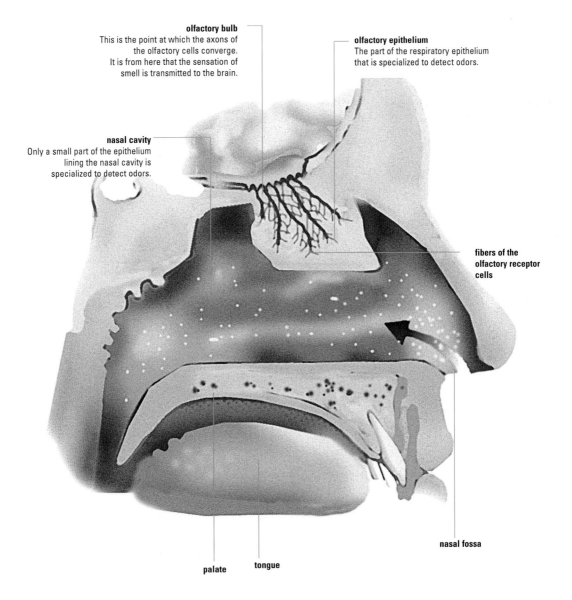

olfactory bulb
This is the point at which the axons of the olfactory cells converge. It is from here that the sensation of smell is transmitted to the brain.

olfactory epithelium
The part of the respiratory epithelium that is specialized to detect odors.

nasal cavity
Only a small part of the epithelium lining the nasal cavity is specialized to detect odors.

fibers of the olfactory receptor cells

nasal fossa

palate

tongue

The systems of the human body

The senses

Taste

The structures that receive and transmit the sense of taste are known as taste buds. Most are situated on the papillae of the tongue, although some are also found in the pharynx, the palate, and elsewhere. Each taste bud consists of taste receptor cells enclosed within supportive tissue, and opens into the lateral canal of the papilla, known as the pore canal. It is here that the cilia of these specialized cells come into contact with the substances, dissolved in the saliva, that produce taste.

There are four basic types of taste: sweet, sour, bitter, and salt. Each is identified principally with a specific area of the tongue. Many different sensations are possible through combinations of these tastes, together with other associated stimuli such as odors (see illustration on page 169).

The sensations of taste and smell are closely associated. Conditions that adversely affect the sense of smell also result in a diminished ability to taste.

Taste bud

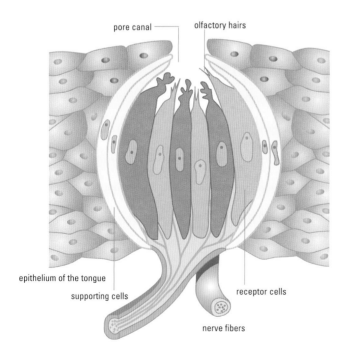

pore canal — olfactory hairs

epithelium of the tongue
supporting cells
nerve fibers
receptor cells

Anatomy

View of the tongue showing the taste regions

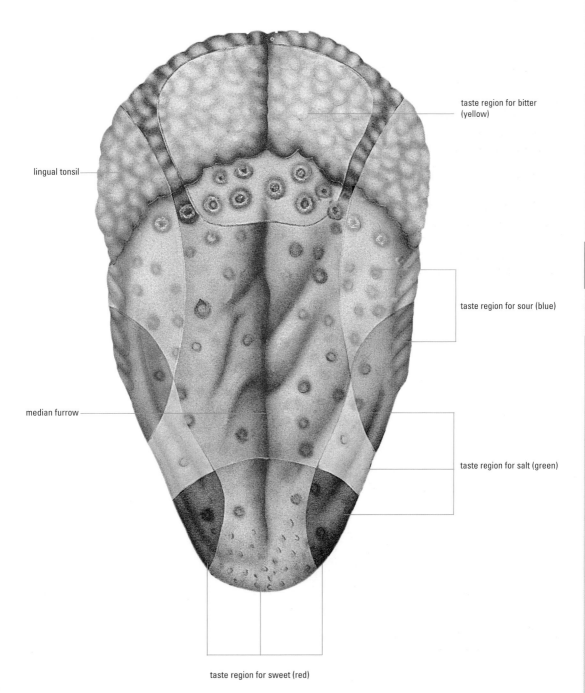

The systems of the human body
The senses

Touch

Touch is experienced through sensory receptors located in the skin or in the subcutaneous tissue. From these receptors, nerve impulses travel to the brain where the appropriate response, voluntary or involuntary, is initiated. There are different types of receptor; some are enclosed within a capsule of connective tissue while others are exposed. The Meissner corpuscles form part of the first category and play a part in the detection of light tactile stimuli, while the Pacinian corpuscles, which are larger, detect stronger tactile stimuli, pressure, and vibration. In addition, free nerve endings are widely distributed throughout the surface of the skin and are sensitive to light touch, pressure, pain, and temperature.

The skin also detects the sensations of heat and cold by means of thermoreceptors. An automatic, involuntary response is triggered by such sensations. In the case of cold this response is aimed at preserving heat and in the case of heat the reverse is true. The diagram below illustrates these processes.

The brain interprets the information picked up by the different sense receptors and responds appropriately.

Phases in the process of thermoregulation

Response to heat

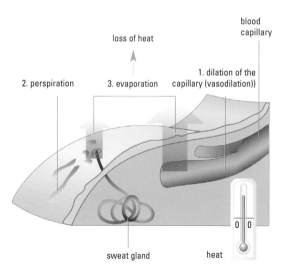

Response to cold

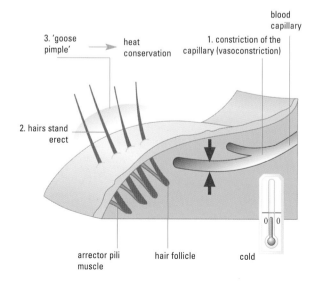

Structure

Structure of the skin and touch receptors

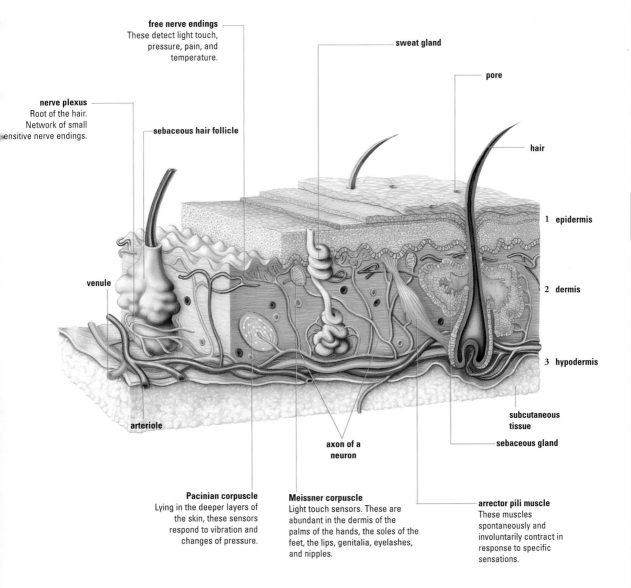

Bibliography

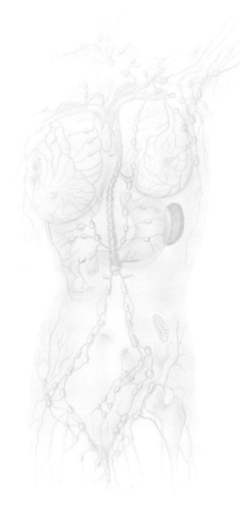

BANNISTER, L.H. 1995: *Gray's Anatomy*. Philadelphia, Churchill Livingstone.

GUYTON, A.C., HALL, J.E. 2000: *Textbook of Medical Physiology*. Philadelphia, W.B. Saunders Co.

KUMAR, V., COTRAN, R.S., ROBBINS, S.L. 1997: *Basic Pathology*. Philadelphia, W.B. Saunders Co.

NETTER, F.H. 1965: *Endocrine System and Selected Metabolic Diseases*. Rochester, Novartis Medical Education.

NETTER, F.H. 1986: *Nervous System*. Rochester, Novartis Medical Education.

Guías Atrium de Medicina y Salud. 1999, Barcelona, Atrium Internacional.